A Specialist Periodical Report

Radiochemistry
Volume 3

A Review of the Literature Published during 1974 and 1975

Senior Reporter
G. W. A. Newton, *Department of Chemistry, University of Manchester*

Reporters
B. W. Fox, *Christie Hospital, Manchester*
G. R. Harbottle, *Brookhaven National Laboratory, New York*
J. A. Heslop, *I.C.I. Physics and Radioisotope Service, Billingham*
D. Sylvester, *Hammersmith Hospital, London*

© Copyright 1976

The Chemical Society
Burlington House, London, W1V 0BN

ISBN: 0 85186 274 8

ISSN 0301-0716

Library of Congress Catalog Card No. 72-92546

Printed in Great Britain
at the Alden Press, Oxford

Foreword

The general design of this volume is different from the previous two in that it is concerned entirely with some of the applications of radioisotopes, one chapter being devoted to industrial uses, one chapter to archaeology, and two chapters to some utilizations of medical importance. It has been the intention to give a comprehensive cover of the literature from 1973 to late 1975 or early 1976, and the Reporters apologize for any significant omissions or misrepresentation.

G.W.A.N.

Contents

Chapter 1 Industrial Applications of Radioisotopes
By J. A. Heslop

1 Radiotracers	1
Flow Measurements	3
Leak Detection	6
Material Movement and Residence-time Distributions	6
Residence-time Distributions	7
Mass-balance and Chemical Reaction Studies	8
Wear	9
2 Sealed-source Techniques	10
γ-Ray Absorption	10
Level and Interface Measurements	11
Density and Particle-size Measurements	12
Thickness Gauges	15
3 Analytical Applications of Radiometric Instruments	17
γ-Ray Absorption	17
Radioisotope-induced X-Ray Fluorescence	17
Backscatter Analysis	20
Neutron Methods	21
Neutron Sources	21
Neutron Absorption	22
Neutron Thermalization	23
Neutron Activation Analysis	25
Well Logging	30
Radiography	31
γ-Radiography	31
Neutron Radiography	32

Chapter 2 Activation Analysis in Archaeology
By G. Harbottle

1 Introduction	33
2 Nuclear Activation of Archaeological Materials	36
3 Interpretation of Analytical Data in Provenience Studies	42
The Data Matrix and its Transformations	43
The Hypergeometric Distance Matrix	46
Cluster Analysis	49

Single Linkage	49
Complete Linkage	50
Average Linkage and Centroid	50
Size-of-Cluster Methods	50
Iterative Clustering	51
The Presentation of Cluster Analysis Results: Dendrograms	52
Correlation Effects, Group Properties, and Mahalanobis Distance	52
Principal Components Analysis	58
Computer Programs	59
Summary	60

4 Application of Nuclear Techniques to Specific Classes of Archaeological Material — 60

Ceramic Provenience Studies	61
Ancient Glass and Faience	65
Obsidian	66
Precious Metals and Coins	66
Turquoise, Jade, and Amber	68
Other Stones of Archaeological Interest: Sandstone, Steatite, Sanukite, Marble, and Flint	69
Steel, Bronze, Copper, and Lead	70

5 Conclusions — 71

Chapter 3 Preparation of Radiopharmaceuticals and Labelled Compounds using Short-lived Radionuclides
By D. J. Silvester

1 Introduction — 73

2 Sources of Short-lived Radionuclides — 74

Nuclear Reactors	75
Charged-particle Accelerators	75
Radionuclide-generator Systems	76

3 Carbon-11 — 76

Syntheses with $^{11}CO_2$	78
Syntheses with $H^{11}CN$	82

4 Nitrogen-13 — 82

5 Oxygen-15 — 85

6 Fluorine-18	85
^{18}F Intermediates	87
Preparation of Specific ^{18}F-Organic Compounds	88
7 Bromine-77	90
8 Iodine-123	91
Direct Production of ^{123}I	91
Indirect Production of ^{123}I	92
^{123}I-Labelled Compounds	93
9 Astatine-211	96
10 Other Radionuclides	97
Gallium-67 and -68	98
Indium-111 and -113m	100
Technetium-99m	103
Lead-203	105
Miscellaneous Radionuclides	107

Chapter 4 Sample Preparation Procedures for Liquid Scintillation Counting
By B. W. Fox

1 Introduction	108
2 Basic Design of Instrumentation and Scintillation Vials	109
3 Solvents and Solutes in the Scintillator System	110
4 Homogeneity and Heterogeneity	112
5 Preprocessing for Homogeneous Liquid Scintillation Counting	112
6 Inorganic Ions	113
7 Combustion Methods	114
8 Solubilization Methods	115
9 Cerenkov Light Measurement	116
10 Preprocessing for Heterogeneous Liquid Scintillation Counting	117
11 Monitoring of Continuous-flow Systems	118

12 Scintillant in the Liquid Phase	118
13 Colloidal Scintillation Counting	121
14 Correction Procedures, Artefacts, and Data Processing	122
15 Chemiluminescence and Phosphorescence	125
16 Biochemical Applications	126
17 Radioimmunoassay and Related Techniques	128
18 Archaeological and Hydrological Applications	130
19 Miscellaneous Applications	131

Author Index 133

1
Industrial Applications of Radioisotopes

BY J. A. HESLOP

The application of radioisotopes to the study of industrial process and to the measurement and control of those processes is well established. This review will contain little that is radically novel in terms of basic techniques or applications. There exists, however, a basic information gap between the radioisotope applications specialist and the user, be he chemical engineer, R and D scientist, or civil engineer. This gap causes the use of radioisotopes as tracers or in instruments to be considered as a last-gasp effort that is to be used only when all else fails. Experience has shown that the use of radioisotopes can often solve problems much more easily than conventional techniques, and, in cases where the radioisotope specialists are part of the industry or belong to an institute with good industrial contacts, the number of radioisotope applications can show an amazing growth (*e.g.* the group at ICI Petrochemicals Division who are concerned with the application of radioisotopes within industry complete well in excess of 1000 applications per year). It is hoped that this review will help to bridge the information gap in that it covers recent applications of radioisotopes in industry up to about mid-1975.

The field has already been the subject of a number of general reviews which cover the basic principles and certain applications.[1-4] Specific reviews covering applications in the textile and fibres industry,[5-10] in plastics,[11-13] and in the basic metals industry[14] have been published.

1 Radiotracers[15]

It has been said that there is no such thing as the perfect tracer, as this material

---

[1] E. W. Stene, J. H. B. George, and H. P. Beutner, Isotopes in Industry Contract AT (30-1)-3337, United States Atomic Energy Commission.
[2] L. J. Taylor, *Reports Progr. Appl. Chem.*, No. 19, p. 673.
[3] H. Simpson, *Reports Progr. Appl. Chem.*, No. 19, p. 243.
[4] J. S. Charlton, J. A. Heslop, and P. Johnson, *Physics in Technol.*, 1975, 67.
[5] W. Markiewicz, *Przeglad Wlokienniczy*, 1973, **27**, 299.
[6] J. Luenonschloss, (Eurisotop-77) Commission of the European Communities, Brussels, 1973.
[7] H. Beckstein, *Chemiefasern Text-Anwendungstech./Text. Ind.*, 1973, **23**, 36.
[8] R. Merkle, *Izotoptechnika*, 1973, **16**, 593.
[9] R. Merkle, *Chem.-Anlagen Verfahren*, 1973, **164**, 71.
[10] V. S. Akopov, N. Yu. Myrin, and V. I. Postnikov, 'Methods for the Technical Economies: Analysis of the use of Radioisotopic Control Methods in Industry', Atomizdat, Moscow, 1974.
[11] A. Kosmowski, *Konstr. Elem. Methods*, 1973, No. 4, p. 113.
[12] L. Trajkov, *Khim. Ind. (Sofia)*, 1972, **44**, No. 5, p. 220.
[13] B. Ya. Munkov, *Trudy Vses. Nauch.-Issled. Inst. Gidrotekh. Melior.*, 1970, **48**, 8.
[14] 'Nuclear Techniques in the Basic Metal Industries,' Proceedings of Symposium at Helsinki, July 1972, IAEA, Vienna, 1973.
[15] C. W. Sheppard, 'Basic Principles of the Tracer Method', Wiley, New York, 1962.

would be indistinguishable from the population as a whole. Isotopes usually provide the nearest approach to the perfect tracer, in that they are, to a good first approximation, chemically and physically indistinguishable from the total population. If they are also radioactive then they possess a useful property (the emission of radiation) which can be related to their concentration in the bulk of the medium under investigation. There are several degrees of sophistication possible in the selection of a tracer. If the medium under investigation undergoes a chemical reaction or a phase change, then a tracer which parallels this behaviour must be selected, and hence the chemical form of the tracer must be identical with the bulk material. If this criterion can be satisfied, then, neglecting any possible isotope effects, the ideal tracer will be used.

In the chemical industry, the isotopes which come into this category are usually ^{14}C and ^{3}H, both of which can be readily incorporated into organic materials. These isotopes are far from ideal for use in industry, in that they are both weak β-emitters and hence cannot easily be detected inside process equipment, and sampling is necessary. They both have long half-lives and hence cannot easily be used to study problems in which they would end up in a product which would be sold to the general public. Despite these disadvantages, there are certain problems which can only be solved by the use of ^{14}C or ^{3}H, and, provided the financial incentive is large enough, both isotopes can be used on full-scale plant both efficiently and safely. On smaller scale equipment and in the laboratory ^{14}C and ^{3}H play very important parts in industrial monitoring.

In a majority of situations, where the bulk property of a material is being followed, there are no physical or chemical changes, and it is possible to use a physical tracer. This type of tracer will follow the bulk movement of a material, provided that it does not undergo a phase change or a chemical reaction. Similarly, the tracer itself must not be precipitated or in any way lost from the material being traced. It is often sufficient if the material is simply soluble in the medium, as in the use of $NH_4{}^{82}Br$ for the tracing of aqueous streams. The use of a physical tracer allows a γ-radiation detector to be used outside the process vessel, and hence removes the need for samples to be taken. This means that short-lived isotopes can be used, thus reducing the problems of radiological protection. Table 1[16—27] is a list of a number of isotopes which have been used in the study of various industrial problems. The isotopes are in general produced in a nuclear reactor and are of sufficiently long half-life to allow easy transport to the industrial site. Exceptions to this are ^{41}A and ^{56}Mn, whose half-life requires reasonably rapid access to an isotope production facility. The production of isotope generators for use in medicine has not been widely applied in industry, usually because the γ-energy requirements are different, although

[16] M. Brown, *Internat. J. Applied Radiation Isotopes*, 1974, **25**, 289.
[17] H. H. Gomez, O. Cuello, and S. Rey, *Nuclear Sci. Abs.*, 1974, **29**, 18 546.
[18] E. Garcia Agudo, U. Dante, T. Ohara, and W. Sanchez, *Nuclear Sci. Abs.*, 1975, **31**, 664.
[19] K. Runge and G. Grahl, *Isotopenpraxis*, 1974, **10**, 133.
[20] L. Riedlmayer, J. Riesing, and V. Muehldorf, *Nuclear Sci. Abs.*, 1975, **31**, 14 259.
[21] J. L. Boutaine, Boeing-747 Specialists Meeting, Everett, Washington, U.S.A., February 1974.
[22] B. A. Fries, (Chevron Research Co.) U.S.P. 3 809 898, 7 May 1974.
[23] W. J. McCabe, K. P. Pohl, and O. J. Rowse, *Nuclear Sci. Abs.*, 1975, **31**, 14 257.
[24] K. Krishnamurthy and S. M. Rao, *J. Hydrol.*, 1973, **19**, 189.
[25] Y. S. Kim and B. H. Lee, *J. Korean Nuclear Soc.*, 1974, **6**, 231.
[26] J. Lontiadis and C. Dimitroulas, *Nuclear Sci. Abs.*, 1974, **29**, 24 206.
[27] B. Gorski, C. Beyer, and H. Ulrich, *Isotopenpraxis*, 1973, **9**, 282.

Industrial Applications of Radioisotopes

Table 1 *Radiotracers for process studies*

Isotope	Half-life	Emission used	Chemical form	Medium
^{56}Mn	2.6 h	γ	Acetate[4]	Aqueous
			Naphthenate[4]	Organic
^{24}Na	15 h	γ	Carbonate[4, 16, 17]	Aqueous
			Naphthenate[4]	Organic
			Salicylate[4]	Organic
^{82}Br	36 h	γ(var 0.55–1.48 MeV)	Ammonium bromide[4, 18]	Aqueous
			KBr[4]	Aqueous
			p-dibromobenzene[4]	Organic
			Methyl bromide[4]	Gas
			Ethylene dibromide[20]	Gas/liquid
^{140}La	40 h	γ	Oxide[4, 17]	Solid
			Acetate[4, 19]	Aqueous
			Naphthenate[4]	Organic
^{198}Au	2.7 d	γ	Colloidal[4]	Organic
^{122}Sb	2.8 d	γ	Sb_2O_3[4]	Solid
^{41}A	110 min	γ(1.29 MeV)	Element[4, 16]	Gas
^{125}Xe	18 h	γ	Element[4]	Gas
^{133}Xe	5.3 d	γ(81 keV)	Element[4, 21]	Gas
^{85}Kr	10.6 y	γ(0.51 MeV)	Element[22, 23]	Gas
^{46}Sc	84 d	γ	Glass[24]	Solid
^{192}Ir	74 d	γ	Glass[24]	Solid
^{51}Cr	27.8 d	γ	edta complex[25, 26]	Aqueous
^{3}H	12 y	β	Depends on chemical form of system under investigation[27]	
^{14}C	5.7×10^3 y	β		

a ^{140}Ba/^{140}La generator has been successfully developed and used by Runge and Grahl[19] to investigate matter-transport processes in chemical plants.

Flow Measurements.[31]—Most industrial processes require measurements of mass flow in order that the efficiency of their operations can be monitored. A large number of conventional flowmeters exist but they often need calibration, or the measurement of an unmetered flow is required. Radioisotope methods have achieved a wide acceptance, often as standard methods of flow measurements where high accuracy is required, for example in the measurement of a mass balance on a plant.

There are several radiotracer methods of flow measurement in common use. The simplest is the pulse-velocity measurement, in which a sharp pulse of radioactivity is injected into the stream whose flow is to be measured. The passage of the pulse is observed by a pair of detectors positioned downstream of the injection pipe at a distance such that lateral mixing within the pipe is complete. The flow rate is given by $Q = lA/t$, where Q is the volume flow rate, A the cross-sectional area, l is the distance between the two detectors and t is the time taken for the radioactivity to travel the distance l. If a γ-emitting isotope is used then the detectors can be outside the pipe and can be connected to either simple or integrating ratemeters. In both cases the time t is obtained by measuring the time interval between the peak half-height positions. The method requires turbulent flow in the pipe and a knowledge of the internal pipe diameter, which must be effectively constant. This latter measurement can be obtained by measuring the external diameter of the pipe and then the pipe wall thickness, using, for example, an ultrasonic method. If accurate flow rates are required this is done at several points along the pipe length, and measurements are also made of the pressure (for gases) and temperature within the pipe.

Under ideal conditions, Evans et al.[28] measured the flow rate of air along a pipe, using ^{85}Kr at flow rates varying from 3 to 300 l s^{-1}. Over 90% of a total of 61 tests carried out had a mean deviation of $< \pm 0.4$% from the flow rate determined by collection of the gas over a given time interval.

Under industrial conditions it is difficult to achieve this sort of accuracy, but deviations of ± 1—2% are usually easily achievable, and under conditions in which the pulse-velocity technique can be used it will be the method of choice for flow measurement.

Flows of liquids and gases can also be measured by a dilution technique in which a radioactive tracer (radioactive concentration S_1) is continuously injected at a rate q into the material flow. Samples are taken at a suitable distance from the injection point. If S_2 is the radioactive concentration at the sample point and Q is the flow rate to be measured, then

$$qS_1 = (Q+q) S_2 \quad (1)$$
$$Q = q(S_1 - S_2)/S_2 \quad (2)$$

In general, $S_1 \gg S_2 .\therefore$

$$Q \approx q\, S_1/S_2 \quad (3)$$

Provided that turbulent flow exists, the measurement of flow rate is independent of the vessel diameter and any changes in it, and hence the method can be used in situations where the pulse-velocity method cannot be applied. Injection is continued until a radioactivity 'plateau' is obtained at the sampling point, and this can be measured with good precision. The radioactive concentration of the injected material and the rate of injection can be accurately measured, and hence overall accuracies of ± 1% can be attained.

The accuracy of the method has been tested against direct weighing by measuring the flow rate along a pipe to a road tanker by the dilution method and comparing this with the weight of material found in the tanker. The results[29] shown in Table 2 show the excellent agreement obtained between the direct method and the dilution technique. This method has been applied by Clayton and Evans[30] to the measurement of flow through turbines and pumps in power stations and is in routine use throughout ICI for the measurement of a wide variety of gas and liquid flow rates. Gas flow rates in excess of 10^5 m^3 h^{-1} and liquid flow rates greater than 10^6 gallon h^{-1} have been measured.

A third method of flow measurement exists which is useful for the measurement of large flows in open channels.[31] The dilution sudden-injection method has a number of variations but essentially consists of the injection of a suitable tracer for a short duration, followed by downstream sampling over a period of time sufficiently long to ensure that the whole of the tracer has passed the sampling point.

[28] G. V. Evans, R. Spackman, M. A. J. Aston, and C. G. Clayton, in 'Modern Developments in Flow Measurement', ed C. G. Clayton, Peter Peregrinus Ltd., London, 1972, p. 245.
[29] P. Johnson and J. Whiston, unpublished work.
[30] C. G. Clayton and G. V. Evans, in ref. 28, p. 276.
[31] International Standard; Ref. No. ISO 555/11-1974 (E).

Table 2 *Measurement of liquid flow by the radioisotope dilution method. A test of the accuracy of the method compared with direct weighing.*[29]

Sample time/min	Radioactive concentration/ counts s^{-1} ml^{-1}	Instantaneous flow rate/ton min^{-1}
2	0.330	0.355
4	0.331	0.354
6	0.331	0.354
8	0.334	0.351
10	0.336	0.350
12	0.335	0.350
14	0.335	0.350
16	0.337	0.348
18	0.338	0.348
20	0.337	0.349
22	0.341	0.344
24	0.340	0.345

Radioactive concentration of injection solution = 6.93×10^4 counts $s^{-1} ml^{-1}$; rate of injection = 19.74 ml min^{-1}; time of addition = 24 min 30 s.
Weight of material as determined by radioisotope dilution flow
  = 8.61 ± 0.06 ton.
Weight of material by direct weighing   = 8.632 ton.

Then

$$S_1 V = Q \int_0^t t S_2 \, dt \qquad (4)$$

giving $Q = S_1 VF/N$ (Total count method) or $Q = S_1 VF/\bar{r}t$ (Continuous Sample method), where F is the efficiency of the counting set-up, S_1 is the specific activity of the injection solution, S_2 are the specific activities of the samples, Q is the volume flow rate, t is the time of sampling after tracer injection, N is the total number of counts accumulated, and $\bar{r}$ is the counting rate for a homogenized sample.[32] The flow rate can be determined either by sampling (total sample method) or simply by using a ratemeter and obtaining a count *versus* time trace at the sampling position (total count method). Neither method is as accurate as the dilution flow, but they are convenient in the measurement of large flows in open channels, *e.g.* effluent flow, as the amount of radioactivity required is much less than that used for the dilution flow method.[18]

Radioisotope measurements can provide an instantaneous method for the measurement of flow with an ease and accuracy that is difficult to achieve by other methods. The technique is widely used within the oil[33] and chemical industries, both for the measurement of unmetered flows and for the calibration of existing flowmeters. The checking of material balances within chemical plants usually depends ultimately on the accuracy of measurement of the flow rates of the materials entering and leaving the plant. In recent years the greatly increased legislation concerned with environmental conditions has led to a greatly increased demand for the measurement of flow in drains and open channels.

---

[32] K. Ljunggren, Symposium on Radioisotope Tracers in Industry and Geophysics, IAEA, Prague, 1966, p. 303.
[33] D. F. Rhodes, *Instrumentation Technol.*, 1975, October, p. 43.

Leak Detection.[34]—The ease of detection of radioactivity and the unambiguous nature of the qualitative determination of its presence make the use of radiotracers ideal for the detection and measurement of leaks. The methods used depend on the system under investigation but the techniques have been applied in situations ranging from underground pipelines[20] to heat-exchanger shells,[23] reactor cooling systems, and aircraft tyres.[21] Leaks are often detected by the continuous injection of a radioactive tracer until the section within which the leak is suspected is uniformly labelled. The tracer within the pipe is then flushed away and a survey is conducted to detect any residual activity in the area adjacent to the pipeline or vessel which has escaped through leakage. It is often advantageous if, before flushing the pipeline to remove radioactivity, a valve is closed and the pipeline is pressurized. Detection of the residual activity from the leak may be carried out by the use of sensitive detectors by traversing the pipeline length or by the use of a pig containing a radioactivity detector which traverses the line within the pipe[35] and prints out the distance from the starting point whenever the radioactivity exceeds a predetermined threshold.

In chemical plants, which often involve a number of process units and heat exchangers, often the first indication of leakage problems is the production of off-specification material. It is obviously very desirable to isolate any leakage, so that the plant down-time may be minimized or avoided by the use of alternative equipment. In situations like this the leaks must be detected while the plant is on-line, and the technique generally used is the injection of a pulse of activity into the material (often cooling water) that is thought to be leaking followed by sampling of the product. This is carried out at a number of points until the position of the leakage is known to be within a single unit. Physical tracers can often be used for this type of study, e.g. ^{24}Na for the detection of leaks of cooling water, but care must be exercised in some cases where leakage may take place from the liquid or the gas phases, *e.g.* leaks of steam into process material, where negative results have been obtained using both a liquid (^{24}Na) and gaseous (^{41}A) tracer but a significant leak has been detected when a chemical tracer is used (^3H_2O). In this case it is obviously necessary that the tracer be compatible with both systems, *i.e.* the leaking material and the product material, a condition which is difficult to achieve with physical tracers.

Material Movement and Residence-time Distributions.—The way in which material moves through a processing unit must obviously have a profound effect on the quality of the final product obtained. Thus a study of the mode of matter transport within a given unit may greatly influence the mode of operation of that unit and also the design of any new units.

The simplest application of radioisotope techniques in this type of study is in the measurement of mixing times in a simple batch process. In this case a pulse of radioactivity is introduced into the mixing vessel and then samples are taken at intervals until the distribution of radioactivity within the system is uniform. The determination of mixing times obviously allows the most effective use to be made of expensive mixing equipment.

[34] M. Gerrard, *Isotopes and Radiation Technol.*, 1968, **6**, 443.
[35] G. Dorgebray (Elf Union) U.S. P. 3 778 613, 11th December 1973.

Residence-time Distributions. Applications of tracers in the assessment of mixing are trivial compared with the information that can be obtained by injection of a pulse of material into the process stream followed by monitoring of the pulse at a later stage in the process. The shape of the pulse may be analysed so that information is obtained about the movement of material in this part of the process.

In the chemical industry one frequently used model of the process system is that of stirred tanks or pots. The exit pulse is analysed in terms of the mean residence time (t_m) and the number of perfectly stirred pots (n) in a cascade which would produce the observed exit trace. For a closed system into which a pulse of radioactivity is injected (Δ function) the shape of the exit trace can be expressed as:

$$E(t) = \left[\frac{n}{t_m - t_1}\right]^n \left[\frac{(t - t_1)^{n-1}}{\Gamma(n)}\right] e^{-n[(t-t_1)/(t_m - t_1)]} \quad (5)$$

where $\Gamma(n)$ = Gamma function, t_1 = delay time, t = real time.

This simple analysis can be used where the data obtained are fairly crude. With modern injection and detectors the data are usually good enough to allow further analysis in terms of the moments of the residence-time probability density, E, of the tracer, derived from the moments of the input and output traces. In this analysis the input trace is not necessarily a Δ function. The various moments can be analysed for various situations,[36] *e.g.* open and closed pipes with axial diffusion, closed apparatus with gamma distribution and delay, and adsorption processes.[37] The analysis applied to the output curves depends on the information required, and various attempts have been made to gain detailed information from the tracers. Niemi[38] used a particulate ^{59}Fe tracer to study flotation processes in the metallurgical industry. The exit pulse is analysed by the method of least squares. An ideal injection impulse is assumed and the results are used to predict the behaviour of particles under varying conditions of process loading.

A simpler approach has been adopted in the determination of the residence times and the degree of mixing of coke and limestone in iron furnaces.[39] In the study of material movement of solids the choice of tracer is very important, as the behaviour of the material can depend on a number of parameters, *e.g.* density, particle size, and particle strength. The value of using reactor-irradiated process material has been demonstrated by a study on a ferrochrome smelter,[40,41] where the use of activated process material revealed that the pellets of chrome concentrate, the lump ore, the coke, the dolomite, and the quartz feeds to the preheater kiln all had different residence times, depending on the grain size of the particles. The knowledge of the residence time allowed several process developments to be carried out, resulting in a decrease in the amount of unusable product.

The continuous transport of material in the WORCA steelmaking process has been studied, using radioactive tracers [^{193}Au (steel-tracer) and ^{140}La (slag)] to follow the counter-current gravity flow of steel and slag.[42] The process was treated

---

[36] R. E. Goddard, personal communication.
[37] E. Kučera, *J. Chromatog.*, 1965, **19**, 237.
[38] A. Niemi, ref. 14, p. 131.
[39] J. S. Michalik, Z. Bazaniak, J. Palige, K. Świgón, and M. Radiwan, ref. 14, p. 205.
[40] R. Kuoppamaki, J. Kuiesi, and S. Blomquist, ref. 14, p. 227.
[41] A. Tamminen, Second European Conference of Triga Reactor Users, Pavia, September 1972.
[42] T. A. Engh, L. Hansson, and K. Ljienggren, ref. 14, p. 251.

as a one-dimensional reactor, using the dispersion model.[43] A detailed analysis of the flow is given in terms of the Pedet number of the slag and the possible boundary conditions of the reactor.

Studies have also been made on the transport of matter through rotating centrifuge drums,[44] in the batch milling of gold ore using various ore particle sizes,[45] in revolving furnaces (including the study of dust formation),[46] in rotating cement furnaces,[17] and in the cooling drum of a cement furnace, where again it was shown that variations in particle size have a large effect on the mixing and transport processes.[47] A general discussion on the use of radioactive tracers in the metals industry has been published.[48] The chemical industry makes extensive use of this type of technique, although little has been published.[49] Recent published studies have been carried out on a high-pressure hydrogenation unit (using a [^{14}C]octadecanol tracer),[27] on some continuously operating polycondensation reactors,[19] and in polyethylene production.[50]

Although a large amount of information is obtainable from measurements of residence time, it is rare that the analyses have been carried to the full extent. Difficulties can arise by the use of incorrect assumptions as to the conditions present in the process; for example, most inlet and exit tracers are measured by detectors placed externally to the inlet and exit pipes. Thus, at the inlet, the assumption is usually made that the velocity profile across the pipe is flat and the tracer is adequately dispersed. If these assumptions are correct then the average tracer concentration as measured by the detectors is proportional to the fluid flux. If the velocity profile is not flat, or the tracer is not well dispersed, then the average concentration at the section may be grossly misleading. This is the case with a liquid in laminar flow in a pipe. Similar considerations apply to the exit trace, where again turbulent flow is necessary if an external detector or specific point-sampling device is used.

Mass-balance and Chemical Reaction Studies. If the material under study is undergoing chemical reactions during mixing and transport then it is obviously necessary to use a radioisotopic tracer which is chemically and physically identical to the material under study. The use of this type of tracer, coupled with the techniques described in the previous section, allows the chemical reactions occurring in a given residence time to be studied, and by an extension of these techniques the mass balances for a given element or compound may be determined.

The metallurgical industry has once again been at the forefront in applying these techniques. The use of neutron-activated lead–zinc sinter containing primarily ^{69m}Zn, ^{65}Zn, ^{76}As, and ^{122}Sb allowed the transport and distribution of zinc in the

---

[43] T. A. Engh, C.-E. Grip, L. Hansson, and H. K. Womer, *Jerkont. Annlr.* 1971, **155**, 553.
[44] A. van Dalen, *Polytech. Tijdschr., Procestech.*, 1974, **29**, 9.
[45] D. I. Exall and W. J. Taute, *Nuclear Sci. Abs.*, 1974, **29**, 12 993.
[46] I. Torok, *Izotoptechnika*, 1974, **17**, 400.
[47] H. Roctzer and V. Meuhldorf, *Nuclear Sci. Abs.*, 1975, **31**, 14 260.
[48] Yu. B. Belyaev, 'Use of Radioisotopic Tracers for Investigation into Metallurgic Process', Atomizdat, Moscow, 1972 (*Nuclear Sci. Abs.*, 1975, **31**, 3606).
[49] P. Johnson, R. M. Bullock, and J. Whiston, *Chem. and Ind.*, 1963, **19**, 750.
[50] A. C. Castagnet, C. Czulak, M. Said, T. Chara, S. Nakahira, and R. A. Perablo, *Nuclear Sci. Abs.*, 1974, **30**, 23 590.

zinc phase, lead phase, and slag of an Imperial Smelting furnace to be determined.[51] It was also possible to identify the chemical reactions occurring in the various sections of the furnace and to suggest various means of preventing the re-oxidation of zinc vapour within the furnace. A similar study has been carried out on the distillation and purification of zinc and cadmium during all stages of their production.[52]

The mass balances of various elements (S, P, and Cr) in batch iron-making (open-hearth process) and continuous iron-making (Krupp–Reno process) have been studied.[39] The isotopes are injected as pulses of activity, and an attempt has been made to select a suitable chemical form for the tracer, although this presents a number of difficulties when dealing with complex feed materials such as fuel oil and pig-iron. A general method of dealing with the problems of multi-source and multi-product distributions in determining the mass balances has been presented.

In the determination of mass balances, two approaches are possible: (a) the pulse/total count method, in which a pulse of activity is injected and all product streams are sampled continuously and the total activity in each is determined, or (b) the continuous injection/steady-state method, in which tracer is injected over a period, to give a steady-state tracer concentration in the process, and then the mass balance is determined by sampling the product streams at the steady state and comparing their radioactive concentrations with the rate of injection of radioactivity. In general, the second method gives more accurate results and a smaller number of samples to be processed, but often the continuous nature and size of the process being investigated preclude the use of steady-state methods because the amount of tracer that would be necessary would be too large. Studies using both of these methods have been carried out on mechanisms in the deoxidation of steel[53] and in coking processes.[54]

Mass-balance measurements are of prime importance in the chemical and petrochemical industry, and many problems have been studied using the techniques described above. The most useful isotopes are ^{14}C and ^{3}H, incorporated into labelled compounds. These isotopes present problems in that they are long-lived, low-energy β-emitters. Thus sampling is required, and the use of large quantities of radioactivity often presents problems of disposal. It is usually possible to arrange for the dilution of any radioactive product to acceptable levels, e.g. to below naturally occurring levels, and the use of low-level counting techniques which have been developed in connection with radiocarbon and tritium dating can greatly reduce the amount of isotope required.

Wear.—Wear processes and the means of reducing wear by the use of lubricants *etc.* are of great interest to most industrial concerns. Radioactive tracers and methods of analysis have played their part in the study of wear phenomena, especially in reducing the time required to carry out wear studies.

The most usual method of detecting wear is to activate the component under study either by neutron irradiation or by the use of accelerator-produced particles, e.g.

---

[51] N. Biala, M. Brafman, H. Fik, J. Kierzek, R. Kurek, J. Mrowiec, M. Nowak, and Z. Radzikowski, *J. Metals*, 1973, **25**, 22.
[52] K. Akerman, Eurisotop-84, December 1973 (*Nuclear Sci. Abs.*, 1974, **29**, 26 954).
[53] H. Litterscheidt and D. Lohr, Eurisotop-89, June 1974 (*Nuclear Sci. Abs.*, 1975, **31**, 663).
[54] J. Siewierski, H. Kolaski, and H. Firganek, *Isotopenpraxis*, 1974, **10**, 174.

deuterons.[55, 56] This latter technique is preferred, as only the surface layer (30 to 500 μm) is activated, thus reducing the size of the radioactive source used. The rate of wear is then determined by measurement of the amount of radioactivity transferred to the lubricant[57, 58] or by the measurement of chipping or of the fines produced during the use of the irradiated component.[59]

It is, of course, not always necessary to irradiate the component under investigation provided it contains a tracer which can be identified by the use of a sufficiently sensitive analytical technique. Neutron activation analysis (see below) has been used to measure the wear of tools by analysis of post-cutting chips for tungsten (inherently present in the tool) and europium (deliberately added).[60] Recent work by Tolgyessy on the use of ^{85}Kr clathrate compounds and ^{85}Kr implants is beginning to find application in the industrial field. The method consists either of preparing a clathrate of the compounds under investigation or of implanting the ^{85}Kr into the surface of the component to a known depth.[61] Wear or corrosion is then measured by detecting the rate of release of the radioactive krypton gas. The method has been applied to the measurement of the chemical resistance of poly(methyl methacrylate) to various systems[62] and to the measurement of drill-bit wear, using either a clathrate[63] or a grease[64] containing ^{85}Kr.

2 Sealed-source Techniques

The methods discussed so far have generally involved the use of radioactivity in an unsealed form, and the phenomenon of radioactive emission has been used only as a convenient means of detection and measurement of a tracer. Use can also be made, however, of the properties of the radiation itself, and especially by measuring the attenuation of the radiation as it passes through matter. Radioactive sources for this type of application are termed sealed sources, and are fabricated so that leakage of radioactive material from the source occurs only under extreme conditions.

γ-**Ray Absorption.**—The absorption of *γ*-radiation by matter can be represented by

$$I = I_0 \exp(-\mu_{\text{eff}} \rho x) \qquad (6)$$

equation (6), where I is the transmitted radiation intensity, I_0 the initial intensity, x the thickness of matter traversed, ρ the density of the matter, and μ_{eff} is the mass absorption coefficient. For a given *γ*-ray energy and absorbing medium, μ_{eff} is a constant. The absorption coefficients given refer to narrow-beam conditions, and μ_{eff} will always be lower under practical conditions, owing to scattering of the *γ*-rays. For *γ*-rays of energy greater than 200 keV the absorption is dependent mainly upon the electron density of the material traversed, and therefore it is approximately

[55] T. W. Conlon, *Non-Destructive Testing*, 1974, **7**, 310.
[56] G. Katzenmeir, *Kerntechnik*, 1974, **16**, 152.
[57] V. Reudinger, (AED-CONF-73-586-001) October 1973 (*Nuclear Sci. Abs.*, 1974, **30**, 24 133).
[58] V. Reudinger, *Kerntechnik*, 1974, **16**, 164.
[59] M. El Thawil and W. Sanchez, *Nuclear Sci. Abs.*, 1975, **31**, 22 896.
[60] K. N. Prasad, W. A. Jester, and F. J. Remick, *Nuclear Technol.*, 1974, **24**, 252.
[61] V. Jasenak and J. Tolgyessy, *Radiochem. Radioanalyt. Letters*, 1974, **19**, 267.
[62] M. Harangazo, M. M. Naoum, J. Tolgyessy, and S. Varga, *Radiochem. Radioanalyt. Letters*, 1975, **20**, 295.
[63] B. A. Fries, U.S. P. 3 818 227, 18 June 1974.
[64] B. A. Fries, U.S. P. 3 865 736, February 1975.

proportional to the mass per unit area (kg m^{-2}) of the material in the radiation path, regardless of its composition. For low-energy γ-rays the absorption is very dependent on the atomic number of the absorbing medium, elements of high atomic number being the most effective absorbers.

The attenuation equation [equation (6)] shows that measurement of the attenuation of γ-rays provides a means of measuring thickness and density.

Level and Interface Measurements.—The safe and efficient running of industrial plant usually requires control and measurement of level in a variety of equipment and under a range of conditions, such as feed tanks or hoppers, reaction vessels or product-storage tanks, at various temperatures, pressures, *etc*. The levels of materials, which are difficult to measure using conventional gauges, can usually be easily measured using a radioactive gauge in which both the source and detector are positioned on the outside of the containing vessel. The technique can be used to measure the level and to control materials at high temperatures and pressures; materials which are toxic, corrosive, or viscous can be measured in vessels ranging in diameter from a few centimetres to 25 metres or more and with wall thicknesses up to 20 cm of steel. Level gauges are the most widely used type of radioactive control instrument, and their use has been the subject of a number of recent reviews.[65, 66] A comparison of radioactive level gauges with other types concludes that, despite a higher initial cost, the operation and maintenance costs of the radioactive gauge often mean that they are the most economical solution to a difficult problem of measuring levels.[67]

The design of γ-ray gauges, and especially the calculation of source size, γ-ray energy, the type of detector, and the time-constant for different materials and for different materials and for different vessel wall thicknesses, has been discussed both in terms of fundamental properties of the system[68, 69] and by the use of graphical[70] and nomographical[71, 72] techniques.

The simplest type of level-measuring installation, used in vessels in which the process material is always maintained above or below some critical position, is the level alarm. The apparatus is essentially a collimated point source of γ-rays and a detector which is essentially a 'yes/no' device depending on whether or not the detector is receiving the γ-radiation.[73] For certain applications a continuous indication of level is required over a range of heights in the vessel. There are a number of possible configurations for the apparatus for assessing this. For small heights the source may be positioned above the vessel and the detector below, or *vice versa*.

---

[65] W. H. Kinser, 'Radiometric Gauges, A Bibliography (TID-338)', January 1974 (*Nuclear Sci. Abs.* 1974, **29**, 15 772).
[66] 'Radioisotopic Instruments in the Construction Material Industry', ed. E. M. Lobanov, Atomizdat, Moscow, 1973 (*Nuclear Sci. Abs.*, 1974, **29**, 484).
[67] R. C. Spang, *Instrumentation Technol.*, 1972, **19**, no. 12, p. 33.
[68] I. Traikov and V. Vasilev, *Elektroprom. Priborostr.*, 1972, **7**, 176.
[69] A. Notea and Y. Segal, *Nuclear Technol.*, 1974, **24**, 73.
[70] I. Traikov and V. Vasilev, *Elektroprom. Priborostr.*, 1972, **7**, 217.
[71] V. Zlatarov and R. Ivanov, *Elektroprom. Priborostr.*, 1972, **7**, 254.
[72] V. N. Pozdnikov, V. K. Polkovnikov, I. M. Taksar, and V. A. Yanushkovskii, *Izotopy S.S.S.R.*, 1972, **26**, 3.
[73] V. N. Pozdnikov, O. L. Sazonov, I. M. Taxar, E. R. Tesnavs, V. A. Yanuskovskii, A. P. Gavrilov, V. P. Korkonosov, A. V. Peresypkin, and E. P. Shapovalov, U.S. P. 3 848 131, 12 November 1974.

This method is often used to provide a weight control in the filling of containers of consumer products.[74, 75] Over larger vessel depths, then the radioactive source and detection assembly are conventionally placed on opposite sides of the vessel. The radiation shielding is so designed that the radiation beam has a 20° downward spread that is directed across the vessel. The detection system is also extended to cover the region of interest. As the level inside the vessel rises, the radiation over an increasing length of detector is attenuated, and so the signal from the detector decreases. Feedback systems can be used so that the level position can be controlled. The range of levels measured can be extended by the use of multiple or extended sources of radiation.[76] For vessels in which access to both sides is difficult, or for vessels of very large diameter, the source and detector can be placed on the same side of the vessel, and scattered radiation (backscatter) is used to monitor the level. The applications of backscatter gauges are less numerous than transmission gauges for level measurement as they are unsuitable for vessels with wall thicknesses of more than 1 cm. They are also badly affected by the lay-down of any deposits on the vessel walls.

The use of γ-gauges depends on a change of density at the interface between the level to be detected and the other component in the vessel, for example, water/air or ash/air, but the detection of an interface between two materials of the same or approximately the same density is often required; this cannot easily be done using γ-ray techniques. In cases such as these, neutron sources, which rely on the moderation of fast neutrons by the hydrogen atoms of the medium whose level is to be measured, and the detection of slow neutrons are used. Extended neutron detectors can again provide proportional level control, either in conjunction with an extended neutron source or a series of point sources. Neutron level detectors are usually more expensive than the corresponding γ-ray device but are invaluable for the detection of interfaces between phases of nearly equal densities.

Although the γ-ray and neutron systems so far described have been in terms of permanent installations, portable sources and detectors have been developed for both the γ-ray and neutron devices which allow levels to be measured at any time, without the installation of a permanent device, and over greater distances than with the permanent systems. The systems usually consist of a pair of pulley systems on which the source and detector can be lowered so that the γ-ray or neutron transmission can be measured. The neutron system can also be used in the backscatter mode.

Density and Particle-size Measurements.—The attenuation of γ-rays depends on the density of material through which the radiation is passing, as indicated in equation (6). This property has been used in the construction of instruments for the measurement of density.

The theoretical aspects of the measurement of density by radiometric gauging have been discussed in a number of papers, especially with respect to the errors which can arise from field use,[77] due to incorrect use of the instrument and to

---

[74] F. L. Calhoun, U.S. P. 3 784 827, 8 January 1974.
[75] J. Leroy and J. P. Volat, Fr. P. 2 183 561/A, 10 May 1972.
[76] M. Glaeser and K. P. Emmelmann, *Isotopenpraxis*, 1973, **9**, 318.
[77] R. J. Regunato, *Soil. Sci. Soc. Amer. Proc*, 1964, **38**, 156.

Industrial Applications of Radioisotopes

various other conditions such as streaming, source defects, *etc.*[78] Russian experience in the determination of the densities of polyethylene, boron carbide, serpentine filling, and iron shot has been described.[79] Automatic continuous density determination has been the subject of a symposium of European chemical engineers.[80] Various radiometric instruments have been described, utilizing various configurations of source and detector. Two pairs of alternately operated sources and detectors have been used for the differential measurement of density independently of the sample geometry and the chemical composition of the interfering materials.[81, 82] The most usual isotopes used for density measurement are cobalt-60,[83] caesium-137,[84] and americium-241, all of which exhibit useful γ-ray energies and have reasonably long half-lives. γ-Ray absorption density gauges have been used for the measurement of the densities of soil,[83] grindstone disks,[85] fabrics,[86] and nuclear fuel elements.[87] γ-Ray attenuation has also been used for the determination of the particle size of iron ores, using ^{241}Am,[88] the rate of sedimentation in a fluid, using a ^{147}Pm/Al X-ray source where both source and detector move in a vertical plane,[89] and for the assay of the particle contents of fluids by the attenuation of a γ-source by the particles collecting on a filter pad by incorporating a radioactive isotope into the filter.[90] The on-stream measurement of particle size and size distribution by a variety of radiometric methods has been reviewed.[91]

An extension of the density gauge can be used to determine a mass flow rate along a conveyor belt, *i.e.* nuclear belt weighing.[92a] The apparatus consists of a γ-ray transmission gauge which measures mass per unit belt length and a tachometer which measures belt speed. Signals from the two units are processed to give mass per unit time and integrated to give total mass flow.[92b] A comparison with the electromechanical belt-weighing system for the measurement of iron ore has been produced with regard to ease of installation, calibration, precision and accuracy, and maintenance costs. Accuracies of 0.5—2% are claimed for the nucleonic system, with lower installation and maintenance but higher initial capital costs. Both point sources and extended ^{137}Cs sources have been used, with extended detectors, for the measurement of iron ore and coal fines. The belt weigher gave accuracies varying

---

[78] T. C. Piper, *Nuclear Sci. Abs.*, 1974, **30**, 18 769.
[79] Yu. A. Voropaev, P. I. Dubenskov, V. N. Krasnoshchekov, and D. B. Pozdeev, *Ind. Lab. (U.S.S.R.)*, 1973, **39**, 1769.
[80] D. J. Brown, P. F. Nolan, and E. Rothwell, ACHEMA Congress, Frankfurt am Main, Federal Republic of Germany, 20 June 1973 (*Nuclear Sci. Abs.*, 1975, **31**, 619).
[81] P. Kehler, U.S. P. 3 846 631, 5 November 1974.
[82] P. Kehler, U.S. P. 3 840 746, 8 October 1974.
[83] I. I. Kreyndlin, S. S. Mukhin, V. S. Novikov, and A. A. Pravikov, *Trudy Vses. Nauch.-Issled. Inst. Radiats. Tekh.*, 1972, no. 7, p. 80.
[84] B. Menglekamp, *Electro-Apz.*, 1974, **27**, 132.
[85] B. Balla, *Izotoptechnika*, 1974, **17**, 466.
[86] W. Markiewicz, Z. Misztal, M. Maliewska, T. Spodenkiewicz, A. Wojciecheowski, L. Leonowicz, and K. Mader, *Prace Inst. Wlokien, Lodz*, 1973, **22**, 143.
[87] M. A. Winkler, *Nuclear Sci. Abs.*, 1975, **31**, 16 483.
[88] C. B. Daellenbach, W. M. Mahon, and F. E. Armstrong, *Nuclear Sci. Abs.*, 1975, **31**, 3565.
[89] T. Allen, B. P. 1 327 044, 15 August 1973.
[90] B. A. Fries and C. K. Parker, U.S. P. 3 824 395, 16 July 1974.
[91] N. G. Stanley Wood, *Control Instruments*, 1974, **6**, No. 11, p. 42.
[92] (*a*) I. S. Boyce and J. F. Cameron, ref. 14, p. 155; (*b*) A. C. Hold, D. W. Morgan, and K. F. Williams, *ibid.*, p. 165.

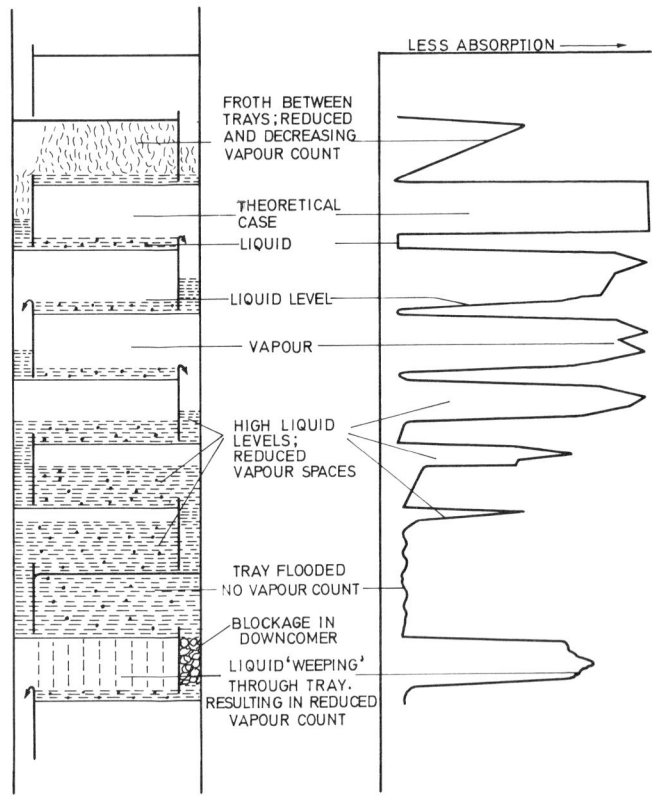

Figure 1

from ±3.2% for ore rubble down to ±0.36% for ore pellets. Some linear extended sources had a non-uniform flux, which led to increased errors.[93]

For materials of lower density, β-ray attenuation may be used. These gauges are especially useful in the measurement of particulate matter in gases.[94,95]

Portable apparatus exists which can be used on installations which do not possess permanent nucleonic installations. This type of equipment is especially useful for the measurement of density profiles on working vessels such as distillation columns. Figure 1 shows a typical absorption profile which allows the easy detection of tray structure, vapour spaces, and column flooding.

Techniques involving the detection of scattering of both β- and γ-rays have also been used in the determination of density. These are especially useful to the determination of surface densities,[96] and a number of theoretical equations which give the distribution of back-scattered radiation with distance from the source for given

[93] P. M. Sigal, *Austral. Process Eng.*, 1973, **1**, No. 5, 25.
[94] H. Dresia and R. Mucha, *Nuclear Sci. Abs.*, 1974, **30**, 24 092.
[95] M. J. Parkinson, U.S. P. 3 766 379, 16 October 1973.
[96] L. S. Pavlov and L. N. Terekhin, *Meas. Tech. (U.S.S.R.)*, 1974, **17**, 259.

Industrial Applications of Radioisotopes 15

source energies and materials have been derived.[97] γ-Ray backscatter has been used for the measurement of the depth of embedding and the spacing of steel cords in rubber conveyor belts.[98]

The scattering of X-radiation has been used to determine the particle size of solid particles of relatively high atomic number suspended in a fluid of relatively low atomic number. The scattered X-rays are of two types, *viz.* Compton-scattered and Rayleigh-scattered, and by suitable choice of the X-ray energy the Compton scattering can be made markedly dependent on the particle size, while the Rayleigh scattering is relatively independent of particle size. Thus comparison of the Compton and Rayleigh scattering provides a measure of the size of the solid particles. Intensity measurements are made using a common detector, with analysing equipment for resolving the energies of the scattered radiation. When absolute mean particle sizes are required, the results are combined with the measurement of solid particle proportion, using a γ-ray density gauge.[99] Particle sizes of solids in the gas phase have been measured using β backscatter and have proved useful in the pharmaceutical industry in the testing of dusting powders.[100] The theoretical aspects of the β-backscatter method of density determination have been discussed.[101]

Thickness Gauges.—Radiometric thickness gauges exist which use a variety of techniques, such as α-, β-, or γ-absorption, β or γ or neutron backscatter, and X-ray fluorescence. The techniques have been the subject of recent reviews,[102—105] and applications in the weaving,[106] textile,[107] and steel[108] industries have been described. The type of nucleonic gauge used depends on the system parameters, such as thickness to be measured and the presence of a substrate or backing material. γ-Ray absorption is used over a wide range of thicknesses by varying the type and the strength of source used. Recent applications include the measurement and control of wide-flanged steel beams, using a single ^{137}Cs source and three separate ionization-chamber detectors. The system is installed in an on-line process-control application and gives an accuracy of measurement of ± 0.1 mm for thicknesses under 10 mm, ± 0.2 mm for thicknesses between 10 and 20 mm, $\pm 1.0\%$ for thicknesses under 90 mm, and $\pm 2\%$ for thicknesses under 120 mm. Control feedback is provided.[108]

The profile of steel sheets can be measured using an ^{241}Am source and an ionization detector that moves across the sheet with a measuring time of less than one minute.[108] Similar systems with a moveable source and fixed extended detectors[109] or with dual sources and detectors[110] exist. The thickness of steel plate has also been

[97] K. Priess and A. Haccoun, *Nuclear Eng. Design*, 1974, **30**, 123.
[98] T. Gregor and L. Misilek, *Jad. Energ.*, 1974, **20**, 186.
[99] (a) K. G. Carr-Brion and S. E. Bramwell, U.S.P. 3 749 910, 31 July 1973; (b) B. P. 1 323 695, 18 July 1973.
[100] I. Jombik and M. Contafalska, *Radiochem. Radioanalyt. Letters*, 1973, **15**, 147.
[101] A. Ott, *Materialprüfung*, 1974, **16**, 132.
[102] (a) K. H. Waechter, *Arch. Tech. Messen Messtech. Praxis*, 1973, **454**, 207; (b) *ibid.*, 1973, **455**, 225.
[103] H. Suzeu, *Tetsu To Hagane*, 1973, **59**, 2011 (*Nuclear Sci. Abs.*, 1974, **30**, 18 770).
[104] A. Kosmowski, *Verfahrenstechnik*, 1973, **7**, 229.
[105] *Stahl Eisen*, 1974, **94**, 818.
[106] J. Trauter, *Chemiefasern Text-Anwendungstech./Text. Ind.*, 1973, **23**, 112.
[107] H. Beckstein, *Chemiefasern Text-Anwendungstech./Text. Ind.*, 1973, **23**, 36.
[108] K. Mijagawa, ref. 14, p. 287.
[109] Y. Murata, Y. Mashiko, Y. Vehida, and M. Matsumoto, U.S. P. 3 868 510, 25 February 1972.
[110] T. Beckage and W. S. Locks, U.S. P. 3 822 383, 2 July 1974.

measured, using a high-energy γ-emitting source and a scintillation detector incorporating a non-linear photomultiplier which provides a linear detector response to changes in thickness.[111] The measurement of large thicknesses of materials of low atomic number has been carried out using monoenergetic X-rays in the 14—30 keV range, produced by a primary γ-source and a detector containing krypton gas, the fluorescent X-rays produced being above or close to the K-edge of the krypton gas.[112]

The lower penetrating power of β-radiation is used for measuring lower thicknesses, and the use of β-gauges for the measurement of plastic sheet has been reviewed.[113] Extremely thin films, with weight per unit area of a few μg cm^{-2}, have been measured using the manganese K-Auger electrons emitted by ^{55}Fe with a G.M. detector.[114] β-Backscatter has also been used for the measurement of films, and especially of coating thickness, where the material of interest is plated onto a backing material of much greater thickness. This method, utilizing a ^{14}C source, has been used for measuring very thin metal films (100—1000 ± 15 Å) produced by vacuum evaporation,[115] for aluminium coatings on plastic substrates,[116,117] for the determination of various metal coating thicknesses on cans,[118] bearing races,[119] and printed-circuit boards,[120] and in the measurement of chromium coatings in the printing industry.[121]

The theoretical aspects of β-backscatter thickness measurements have been discussed in a number of papers.[122—124]

γ-Ray backscatter has been used for the measurement of the thickness of carbon bricks in a blast-furnace lining, using a 10 mCi ^{60}Co source and a scintillation detector coupled to a multichannel analyser. The technique provided a useful alternative to the usual method of measuring erosion of furnace linings by looking for the disappearance of embedded ^{60}Co sources.[125] A differential method of thickness measurement using two parts of the γ-backscatter spectrum has been described.[126]

Radioisotope-induced X-ray fluorescence also provides a useful method for the measurement of coating thicknesses.[127] The coating thickness is determined either by measuring the intensity of the characteristic X-rays induced by γ-ray bombardment of the coating itself or by measurement of the absorption of the characteristic

---

[111] W. G. Bartlett and E. L. Mangan, U.S. P. 3 832 542, 27 August 1974.
[112] B. Y. Cho, E. W. Sturkol, and K. E. Wier, U.S. P. 3 809 903, 7 May 1974.
[113] J. D. Bristow, *Nuclear Sci. Abs.*, 1975, **31**, 6143.
[114] C. Mori, J. Koike, and T. Watanabe, *Nuclear Instr. Methods*, 1974, **121**, 253.
[115] (a) T. Yamashita and T. Mornoi, *Kumamoto Daigaku Kyoikugakubu Kiyo Shizernkagaeu*, 1972, No. 20, pp. 1—6; (b) *Nuclear Sci. Abs.*, 1974, **30**, 18 775.
[116] R. V. Heckman, *Nuclear Sci. Abs.*, 1974, **30**, 29 602.
[117] G. M. Taylor, *Nuclear Sci. Abs.*, 1975, **31**, 6144.
[118] W. Andrae and P. Krautkramer, *Verpackung (Leipzig)*, 1971, **12**, 133.
[119] M. Steeg, *Metalloberflaeche*, 1974, **28**, 127.
[120] H. Fischer, E. I. P. C. Conference on Quality Control of Printed Circuit Boards, Munich, Federal Republic of Germany, August, 1974.
[121] S. Sekowski, *Nukleonika*, 1974, **19**, 365.
[122] L. Pazmandi, P. Tabor, M. Varaljai, and E. Zold, *Isotoptechnika*, 1974, **17**, 155.
[123] B. Martinus, *Belg.-Ned. Tydschr. Oppervlakte* 1974, **18**, 275.
[124] *Adhaesion*, 1972, **16**, 389.
[125] M. Kato, O. Sato, and H. Souto, *Seisan Kenkyu*, 1973, **25**, 541.
[126] J. Seda and L. Musilek, *Nuclear Instrum. Methods*, 1974, **114**, 183.
[127] J. F. Cameron, 'Applications of Low Energy X- and γ-rays', Gordon and Breach, New York and London, 1971, p. 49.

X-rays from the backing material by the coating.[128] The application of X-ray fluorescence (XRF) thickness measurement to Zn, Al, and Sn coatings on steel,[128] to Cr and Sn coatings on steel,[129] to Zn on steel,[130] and to Zn on iron[131] has been described.

3 Analytical Applications of Radiometric Instruments

γ-**Ray Absorption.**—The equation describing the attenuation of γ-rays by matter has already been given [equation (6)]. The mass absorption coefficient μ defined by this equation varies with the effective atomic number of the sample and with the energy of the γ-ray. In general, the mass absorption coefficient decreases with increasing γ-energy and increases in a complex manner with atomic number. This variation of μ with atomic number can be utilized in the analysis of any single element in a matrix of lower atomic number, provided that the composition and density of the matrix remain reasonably constant.

Because of the simplicity of the apparatus, consisting essentially of a source and detector maintained at a fixed distance, this system is very suitable for on-line analysis, and various probes have been described which maintain source and detector at a fixed distance and are of sufficiently rugged construction to be installed directly in the stream to be measured.[132, 133]

Applications involving the measurement of Pb in flotation feeds, using both [137]Cs 225 KeV γ-rays, measuring Pb concentration and slurry density, are in use at Broken Hill, Australia.[134] An accuracy of $\pm 10\%$ of the lead concentration is readily obtainable. Determinations of lead in flotation feed and lead rougher tailing streams have been carried out by measurement of the absorption of [153]Gd γ-rays.[135] The method has also been used for the determination of lead in petrol.[136] A study of the effect of humidity and porosity on the measurement of Cr in ore deposits has been carried out.[137]

Radioisotope-induced X-Ray Fluorescence.—The use of XRF to determine the concentration of selected elements in a variety of matrices has found ready acceptance by most analysts.[138] The conventional techniques involving the use of high-voltage electron tubes and sensitive crystal dispersing units have not found a ready application in the more testing plant environments.

α-, β-, γ-, and X-radiation from a number of radioisotope sources are capable of exciting the characteristic X-rays of most elements. The intensity produced is a factor of 10^6—10^7 less than that available from most X-ray tubes, but the possibility

---

[128] D. K. Danhoffer and C. K. Beswick, 'Nuclear Techniques in the Basic Metal Industries', Proceedings of Symposium at Helsinki, July 1972, IAEA, Vienna, 1973 p. 299.
[129] C. G. Clayton, T. W. Packer, and J. C. Fisher, ref. 128, p. 319.
[130] U. Gruber, *Chemia (Arau)*, 1973, **17**, 344.
[131] O. L. Utt and H. J. Evans, U.S. P. 3 848 125, 12 November 1974.
[132] T. D. Balint and F. Gersey, B. P. 1 339 449, 5 December 1973.
[133] J. S. Watt and W. J. Howarth, ref. 128, p. 105.
[134] W. K. Ellis, R. A. Fookes, J. S. Watt, E. L. Hardy, and C. C. Stewart, *Internat. J. Applied Radiation, Isotopes*, 1967, **18**, 473.
[135] W. K. Ellis, R. A. Fookes, V. L. Gravitis, and J. S. Watt, *Internat. J. Applied Radiation, Isotopes*, 1969, **20**, 691.
[136] K. Jackson and R. B. Pedley, unpublished work.
[137] G. M. Malakhov, A. R. Sotskij, and A. A. Azaryan, *Isotopy S.S.S.R.*, 1973, 15.
[138] K. G. Carr-Brion and K. W. Payne, *Analyst*, 1970, **95**, 977.

Table 3 Industrial and field applications of XRF analysis

Elements to be determined	Sample	Source	Detector	Ref.
Ag, Hg, Au	Alloys	—	—	155
S	Refinery products	—	—	156, 157
Ca, Fe	Limestone slime, iron ore	—	—	152, 158
Rare earths	Rock samples *etc.*	—	—	159—162
Cu	Process streams	^{109}Cd	—	152, 163, 164
Cu	Brass	^{147}Pm/Al	Prop. counter	165
Pu	Mixed oxide fuels	—	—	167
Au, In	Electroplating baths	—	—	168
Sn, W, Pb, Zn, Fe	Process streams	—	—	169
Fe, Si	Process streams	^{109}Cd, ^{55}Fe	Prop. counters	152, 155, 170
Fe, Cu, Zn, Sn, Pb	Process streams	—	Solid state	171
Hg, Be, Pb, Cd, As, V, Mn, Cr, F	Coal, fly ash, fuel oil	—	—	172
Mo	Ores	^{109}Cd, ^{147}Pm, ^{119m}Sn	XeProp. counter	152, 161, 173
Ca, Na	Water	—	—	174
Pb	Air pollution, steel	^{137}Cs	—	152, 175
Pb	Lead brass	^{147}Pm/Al	Scint. counter	165
Pb	—	^{109}Cd	Ge/Li	176
Ca, Fe	Cement raw meal	^{238}Pu	Prop. counter	165
Cr	Chrome ore/tails	^{238}Pu	—	165
Sn	Brass	^{147}Pm/Al	Scint. counter	165
Sn	Process samples	—	—	152
Fe	Concentrates	—	—	152
Nb	Concentrates	—	—	152
Ta	Concentrates	—	—	152
Pb, Zn, Cu	Mineral processing pulps	—	—	177
Cu, Zn, Sn, Pb	Mineral processing streams	^{235}Pu, ^{244}Cu	—	178
V	Steel	^{55}Fe	—	153
Mo	Steel	^{109}Cd	Prop. counter	153
Cu	Alloys	^{241}Am	Li/Si	166

of building compact, robust XRF systems has made radioisotope-induced XRF ideal for instruments designed for on-stream or field analyses.[138] The best radioisotope sources are those which emit low levels of radiation of well-defined energy, as these contribute little or no background radiation at the characteristic γ-ray energies of most elements. The most widely used radioelements are ^{55}Fe, ^{238}Pu, ^{109}Cd, ^{241}Am, and ^{57}Co.[139] Other sources which produce bremmstrahlung radiation can also be used.[140—142]

Several reviews on the use of radioisotope-generated XRF analysis in industrial environments have been published.[133, 143—154] Specific applications of radioisotope-induced XRF in analysis are given in Table 3.[155—178] The choice of geo-

[139] W. A. Swick, *Nuclear Sci. Abs.*, 1974, **30**, 83.
[140] V. Voljin, *Studii ei Cercetari, Fiz*, 1973, **26**, 283.
[141] V. Valkovic, *Contemporary Phys.*, 1973, **14**, 415.
[142] A. Robert, *Izotopenpraxis*, 1973, **9**, 153.
[143] J. S. Watts, *Atomic Energy Austral.*, 1973, **16**, 3.
[144] H. Junzendorf, *Nuclear Sci. Abs.*, 1973, **29**, 9830.
[145] E. P. Leman, 'X-ray Radiometric analysis of Non-ferrous and Rare Metal Deposits', Nedra, Leningrad, 1973.
[146] J. R. Rhodes, 'Energy, Dispersion X-ray Analysis; X-ray and Electron Probe Analyser', ed. J. C. Russ, American Society for Testing and Materials, Philadelphia, 1971.
[147] E. Havronek, A. Bumalova, J. Svitel, E. Dejmokova, M. Herchl, J. Jombik, A. Nemeikova, and O. Cutkova, *Nuclear Sci. Abs.*, 1974, **30**, 20 646.
[148] B. D. Sowerby, U.S. P. 3 780 294, 18 December 1973.
[149] P. Martinelli, Symposium on Nuclear Analytical Techniques in Production and Industrial Use of Noble Metals, Brussels, November 1971.
[150] M. J. Owers and H. L. Shalgosky, *J. Phys.* (*E*), 1974, **7**, 593.
[151] W. J. Howarth, G. J. Wenk, and L. R. Wilkinson, *Canad. Mining Met. Bull.*, 1973, **66**, 76.
[152] K. Papez and J. F. Gameron, ref. 128, p. 21.
[153] C. G. Clayton, T. W. Packer, and J. C. Fisher, ref. 128, p. 319.
[154] E. Murad, *Analyt. Chim. Acta*, 1973, **67**, 37.
[155] R. Cesarco, F. V. Frazzoli, C. Mancini, S. Sauti, and L. Storelli, Symposium on Nuclear Analytical Techniques in Production and Industrial Use of Noble Metals, Brussels, November 1971.
[156] T. B. Rawley, *Nuclear Sci. Abs.*, 1975, **31**, 5569.
[157] B. Dzuimkowski, *Radiochem. Radioanalyt. Letters*, 1974, **16**, 73.
[158] M. Wasilewska, J. Astachowicz, and T. Owsiak, *Nuclear Sci. Abs.*, 1974, **30**, 14 803.
[159] A. Sapakoglu, I. Ozoglu, M. Tan, and H. Erdogan, *Tech. J. Ankara Nuclear Res. Centre*, 1974, **1**, 47.
[160] M. Bonnevie-Svendson and A. Follo, 'Analysis and Application of Rare Earth Materials', ed. O. B. Michelson, Universitetsforlaget, Oslo, 1973.
[161] L. A. Krampit, *App. Metody Rentgenovskogo Anal.*, 1974, No. 13, p. 259.
[162] J. de Neef, *Nuclear Sci. Abs.*, 1973, **29**, 12 526.
[163] E. G. Agudo and M. E. Santos, *Nuclear Sci. Abs.*, 1974, **30**, 23, 551.
[164] A. W. Williams and K. G. Carr-Brown, Industrial Measurement and Control by Radiation Techniques, The Institute of Electrical Engineers, Conference Publ. No. 84, 1972, p. 74.
[165] A. I. Lundan and O. P. Mattila, ref. 128, p. 3.
[166] H. E. Marr, *Nuclear Sci. Abs.*, 1975, **31**, 2832.
[167] J. G. Schnizleiu, T. J. Gerding, and M. J. Steindler, *Nuclear Sci. Abs.*, 1974, **30**, 5944.
[168] B. Halynska, M. Lankosz, A. Markowicz, and M. Wasilawski, *Nuclear Sci. Abs.*, 1974, **30**, 14 838.
[169] G. J. Wenk and L. R. Wilkinson, *AMDEL Bull.*, 1974. 1.
[170] D. J. Reed, J. L. Dalton, and A. H. Gillieson, *X-ray Spectrum*, 1974, **3**, 15.
[171] V. L. Gravitis, R. A. Greig, and J. S. Watt, *Australasian Inst. Mining Met. Proc.*, 1974, **249**, 1.
[172] D. J. von Lehmden, R. H. Jungers, and R. E. Lee, *Analyt. Chem.*, 1974, **46**, 239.
[173] A. P. Ochkur, E. P. Leman, V. V. Kotel'nikov, V. M. Ivonov, and Yu. P. Yanehevskii, *Atomic Energy (U.S.S.R.)*, 1973, **35**, 204.
[174] B. Holymska, *Radiochem. Radioanalyt. Letters*, 1974, **17**, 313.
[175] G. Shani, Israel Nuclear Society Annual Meeting, 26 June 1973.
[176] *Nuclear Sci. Abs.*, 1973, **29**, 9831.

metry,[179, 180] source,[181] and detector for various types of apparatus has been discussed.

Backscatter Analysis.—The backscattering of β-particles can occur by two mechanisms: (i) electron–electron interactions which are approximately independent of atomic number, and (ii) scatter by the nucleus, which increases with increasing atomic number. The total backscattered radiation intensity depends on a complex interaction of these two mechanisms. β-Backscatter occurs only in the surface layer of the sample and has been used to carry out analyses, relying on the relationship that saturation backscatter is roughly proportional to $\sqrt{Z}$.[182]

The analysis of various systems is usually discussed in empirical terms. The analysis of three-component systems has been discussed by Cecal[183] for the analysis of Na^+, Ba^{2+}, and Cu^{2+} and applied by Bulgarian workers to the determination of Ba and Fe in barium ores, the third component being the variable Si,Al,O anionic content. The Ba and Fe were determined by two successive measurements, with and without filters (the filters remove low-energy β-particles), using a 0.1 mCi ^{90}Sr source. Nomographic methods have been used to determine the Ba and Fe contents, with an approximate standard error of $\pm 1\%$, in 10—15 minutes, and the apparatus has been suggested for use in barium ore treatment processes.[184, 185]

The ash content of coals has also been measured by this technique; $^{90}Sr + ^{90}Y$ sources have been used, but 'forward' scattering geometry was found to give a reduced error, associated with the variable iron content of the coal.[186] A gauge for the measurement of the ash content of brown coal has been described.[187]

A more general study of the measurement of the concentration of a number of chemical species in solution has shown that a linear dependence between the intensity of the backscattered β-particles and the concentration of any species exists only over a very limited concentration range. The intensity of the backscattered β-radiation is, however, as expected, proportional to the effective atomic number of the sample solution.[188, 189]

The uses and theories of β-backscatter measurements have been reviewed[190] and the precision and sensitivity of the method have been discussed.[191]

The γ- and X-ray backscatter techniques have also been used, and the theory and uses reviewed.[181] The intensity of backscatter radiation depends on the mass per

---

[177] P. E. Stames and J. W. G. Clark, Radioisotope Instrumentation in Industry and Geophysics, IAEA, Vienna 1966, STI/PVB/112, Vol. 1.
[178] J. S. Watt, R. A. Fookes, and V. L. Gravitis, ref. 128, p. 141.
[179] B. N. Igumnov, R. E. Leonov, and A. E. Trap, *Izvest V.U.Z., Gorn. Zhur.*, 1973, No. 7, p. 149.
[180] Y. Matsui and T. Furuta, *Radioisotopes (Tokyo)*, 1973, **22**, 615.
[181] C. Shenbarg, A. Aladjem, and S. Amiel, *J. Radioanalyt. Chem.*, 1974, **20**, 77.
[182] C. G. Clayton and J. F. Cameron, Radioisotope Instruments in Industry and Geophysics, Proceedings of Symposium at Warsaw, October 1965, IAEA, Vienna, 1966, Vol. 1 p. 15.
[183] A. Cecal, *Kernenergie*, 1974, **17**, 4.
[184] A. Rychvarov and Zh. Karamanova, *Rudodobiv. Met.*, 1972, **27**, 25.
[185] N. Bychvarov and Zh. Karamanova, *Isotopenpraxis*, 1973, **9**, 109.
[186] K. S. Klempner and V. V. Parkhamenko, *Ind. Lab. (U.S.S.R.)*, 1974, **40**, 525.
[187] R. C. Joynt, *Atomic Energy Austral*, 1974, **17**, 9.
[188] G. Marcu and V. Sacelean, 24th IUPAC Congress, Hamburg, Germany, 1973.
[189] G. Marcu and V. Sacelean, *Studia Univ. Babes-Bolyai, Ser. Chem.*, 1974, **19**, 97.
[190] M. Peisach, *Nuclear Sci. Abs.*, 1974, **30**, 18 207.
[191] J. Klas, *Radiochem. Radioanalyt. Letters*, 1974, **19**, 325.

Industrial Applications of Radioisotopes

unit area and the effective atomic number of the sample. The saturated backscatter intensity decreases as the effective atomic number of the sample increases.

Binary systems have been analysed, although not many applications exist. The γ-scattering technique can be used for the analysis of bulk samples, and the penetrating nature of the radiation has been used for the analysis of ores on conveyor belts, using γ-ray resonance scattering. Resonance scattering[192] is an elastic process which occurs when a stable nucleus can absorb an incident γ-ray and reach an unstable excited nuclear state. In regaining its stable state, the excited nucleus emits a γ-ray of the same energy as that absorbed. The process is very selective for the element of interest because most nuclei will absorb only over a very narrow range, and the exciting source chosen generally decays *via* the excited state of the element of interest. Table 4 shows a number of possible isotope pairs.[193] Because of the loss

Table 4 *Possible source/element pairs for analysis by γ-ray resonance scattering*[193]

Element	Source	Half life	γ-Ray energy/MeV
Li	^{7}Be	53 d	0.48
Ti	^{46}Sc	84 d	0.89
Ti	^{48}V	16 d	0.98
V	^{51}Cr	28 d	0.32
Cr	^{52}Mn	5.7 d	1.43
Ni	^{60}Co	5.3 y	1.33
Cu	^{65}Zn	245 d	1.12
Ge	^{74}As	18 d	0.60
As	^{75}Se	120 d	0.27
Cd	^{111}Ag	7.5 d	0.34
Cs	^{133}Ba	10.7 y	0.38
W	^{184}Re	38 d	0.90
Hg	^{198}Au	2.7 d	0.41
Tl	^{203}Hg	47 d	0.28

of energy of the γ-ray on emission from the solid state, gaseous sources are the only practicable ones at present. This severely limits the potential of the technique. $Zn^{65}I_2$, $^{60}CoBr_2$, $^{46}ScCl_3$, and $^{75}SeO_2$ have been used at temperatures of 900, 1000, 1100, and 300 °C, respectively, for the analysis of Cu, Ni, Ti, and As, carried out in the laboratory, and nickel analysis has been carried out on ore on a conveyor belt. Although in the early stages of development, the technique (despite a number of drawbacks) could, because of the specificity of the measurement, have considerable industrial potential. The use of the method for the determination of Cu and Ni in boreholes has been described with analysis times of *ca.* 50 s for 0.5% w/w Cu to $\pm 0.1\%$ w/w. Logging speeds of 0.6 m min^{-1} are possible.[194]

Neutron Methods.—*Neutron Sources.* A number of possible sources of neutron fluxes are summarized in Table 5. The advent of ^{252}Cf and the new generation of neutron generators has greatly extended the range of applications to which neutron methods can be put, and it is to be expected that this will be increasingly reflected

---

[192] B. D. Sowerby, *Nuclear Instr. Methods*, 1971, **94**, 45.
[193] B. D. Sowerby and W. K. Ellis, ref. 128, p. 479.
[194] B. D. Sowerby and W. K. Ellis, *Nuclear Instr. Methods*, 1974, **115**, 511.

in the number of reported industrial applications.[195] The various uses of ^{252}Cf, including a number of industrial ones, have been extensively reported in various issues of *Californium-252 Progress*, published by the U.S. Atomic Energy Commission.

Table 5 *Neutron sources*

Source	Neutron energies available	Flux available
Reactor	Thermal ⟶ Fast	0—10^{16} n cm^{-2} s^{-1}
Isotope ^{241}Am/Be	0—10 MeV	2.2×10^6 n s^{-1} Ci^{-1}
^{238}Pu/Be	0—10 MeV	2.2×10^6 n s^{-1} Ci^{-1}
^{227}Ac/Be	0—10 MeV	1.5×10^7 n s^{-1} Ci^{-1}
^{226}Ra/Be	0—10 MeV	1.3×10^7 n s^{-1} Ci^{-1}
^{242}Cm/Be	0—10 MeV	2.5×10^6 n s^{-1} Ci^{-1}
^{228}Th/Be	0—10 MeV	2×10^7 n s^{-1} Ci^{-1}
^{252}Cf($t_{\frac{1}{2}} = 2.65$ y)	0—8 MeV	2.3×10^9 n s^{-1} mg^{-1} (1 mg = 536 mCi)
Generators	14 MeV	10^{12} n s^{-1}

Neutron Absorption. Some elements have high cross-sections for neutron absorption, and this property can be used for the analysis of these elements. Thus methods have been developed for the measurement of boron and cadmium using a californium source,[196] and for boron, cadmium, and indium in aqueous solutions, using a ^{241}Am/Be source, for continuous process control.[197] Boron (0.1—10%) in bulk boron–aluminium alloys has been measured using a 300 mCi ^{241}Am/Be source, and the method has been used for control laboratory work. Precisions of better than $\pm 1\%$ were obtained.[198] Neutron generators have been used for the determination of small concentrations of boron (0.001%) in steel,[199] using a combined absorption and activation method. The method has also been used for the analysis of boron in solids, slurries, or liquids, using an ^{241}Am/Be source, and giving a precision of $\pm 0.001\%$ B$_2$O$_3$ at the 95% confidence level.[200]

The simplicity of the equipment required for neutron absorption determinations has led to its use in borehole logging, and pulsed neutron–neutron logging is well established;[201] methods for determining Cs,[202] Be,[203, 204] Cl,[205] and Li[206] in rock strata have been developed.

[195] W. C. Reinig, P. H. Permour, and W. R. Cornman, Eleventh Conference on Radioisotopes, Tokyo, November 1973.
[196] G. G. Eicholz and W. A. Hendrix, *Radiochem. Radioanalyt. Letters*, 1974, **19**, 157.
[197] K. D. Hellwig, *Z. analyt. Chem.*, 1974, **268**, 343.
[198] T. B. Pierce, C. R. Boswell, and F. P. Peck, *Analyst*, 1974, **99**, 774.
[199] B. Zitnansky and J. Valent, *Jaderna Energie*, 1973, **19**, 229.
[200] R. B. Pedley and T. Dutson, unpublished work.
[201] V. G. Tseitlin, *Nuclear Sci. Abs.*, 1974, **30**, 321.
[202] A. A. Brem and E. I. Krapivskii, *Ekspress-Inform., Ser. Reg., Razved. Promp. Geofiz.*, 1973, 13.
[203] N. M. Kinchekko, A. V. Gorev, and S. V. Alyavdin, *Ekspress-Inform., Ser. Reg., Razved. Promp. Geofiz.*, 1973, 22.
[204] K. F. Kramich and A. V. Gorev, *Ekspress-Inform., Ser. Reg., Razved. Promp. Geofiz.*, 1973, 27.
[205] R. A. Resvanov, 'Nuclear Geophysical and Geoacoustical Logging of Cased Boreholes in Additional Exploration of Oil and Natural Gas Deposits', Moscow, 1973, p. 125.
[206] A. A. Brem, E. I. Krapivskii, and V. B. Sal'tsevich, *Ekpress-Inform., Ser. Reg., Razved. Promp. Geofiz.*, 1973, 1.

The measurement of the gross thermal neutron absorption cross-section of crushed rock samples, using a pulsed neutron source technique, has been described.[207]

Neutron Thermalization. Most isotope sources and neutron generators are sources of fast neutrons. The neutrons from these sources lose energy by a variety of processes on passing through matter, and eventually become thermalized ($E = 0.025$ eV). Hydrogen, with a nuclear mass approximately equal to that of the neutron, is especially efficacious in slowing down neutrons, and hence with a fast neutron source the thermal neutron flux provides a measure of the hydrogen concentration in the surrounding medium.[208—210] The determination of the moisture content of soils,[211—218] foods,[219] coke,[220] coal,[221] iron ore,[222—224] plastics,[225] and rocks[223] has been described. Several geometrical arrangements of the source and detectors are possible, depending on the sample to be measured, and the choice of geometry for a number of applications in the iron industry has been outlined.[224, 226] Experience obtained during the use of moisture gauges on conveyor belts has been described.[227—229]

Although the neutron thermalization method has so far been described in terms of moisture measurement, it actually gives a measure of hydrogen concentration in the sample.[210] This measurement has been used for the determination of hydrogen to carbon ratios in organic liquids. This ratio is important in some industrial processes, such as olefin production by thermal cracking, where an idea of the severity of the cracking required for a given feedstock can be obtained from the H/C ratio of the material. Figure 2 shows a plot of thermal neutron flux *vs.* hydrogen concentration for a number of organic liquids.[230]

The measurement of hydrogen concentrations in rock strata can provide valuable

---

[207] L. S. Allen and W. R. Mills, Proceedings of the 15th Annual Logging Symposium, McAllen, Texas, U.S.A., June 1974, p. B1.
[208] N. Wadci, *J. Radioanalyt. Chem.*, 1974, **23**, 147.
[209] G. Csom and S. Benedek, *Isotopenpraxis*, 1973, **9**, 284.
[210] N. Wada, *J. Radioanalyt. Chem.*, 1974, **23**, 147.
[211] B. C. Renaud, Thesis. Mexico City University, 1973 (*Nuclear Sci. Abs.*, 1974, **30**, 6769).
[212] N. K. Vaswani, Virginia Highway Research Council, Charlottesville U.S.A., Interim Reports, No. 56, January 1974.
[213] OECD Report, Road Research Group, Paris, December 1973.
[214] J. Merriman, *Nuclear Sci. Abs.*, 1975, **31**, 621.
[215] R. Shelke, *Pr. Zakresu Les*, 1971, **32**, 163.
[216] S. L. Webster, *Nuclear Sci. Abs.*, 1975, **31**, 6139.
[217] P. Petrov, *Pchvozn. Agrokhim.*, 1972, **7**, No. 3, pp. 3–8.
[218] N. Soito and K. Kimura, *Nuclear Sci. Abs.*, 1974, **30**, 3406.
[219] S. Helf, *Nuclear Sci. Abs.*, 1975, **31**, 22 795.
[220] J. F. Cameron, *Nuclear Sci. Abs.*, 1975, **31**, 3564.
[221] A. W. Hall, J. L. Koncheski, and R. F. Stewart, *Nuclear Sci. Abs.*, 1974, **30**, 21 254.
[222] H. P. Dibbs, *C.I.M. Bull.*, 1973, **66**, 61.
[223] A. Kreft, *Nukleonika*, 1973, **18**, 615.
[224] K. Papez, J. F. Cameron, and B. Machaj, ref. 128, p. 183.
[225] G. Sh. Pekarskii and V. B. Elagin, *Ind. Lab.* (*U.S.S.R.*), 1974, **40**, 1003.
[226] F. Griezer and R. Klein, ref. 128, p. 73.
[227] E. Frevert and G. Stehno, ref. 128, p. 193.
[228] Y. M. Chen, U.S. P. 3 794 843, 26 February 1974.
[229] J. M. Eskes, *Canad. Contr. Instrum.*, 1972, **11**, 30.
[230] J. S. Charlton and R. B. Pedley, unpublished work.

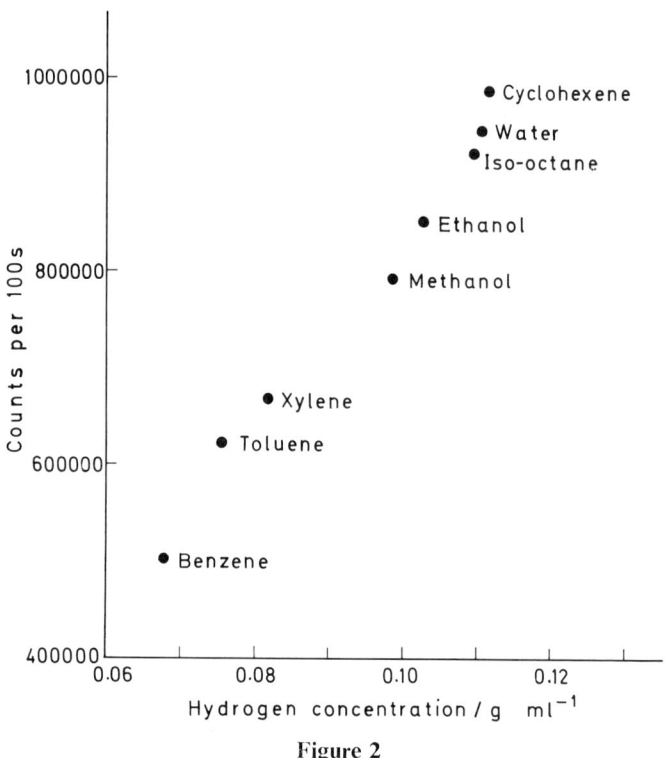

Figure 2

information in prospecting for oil, gas, and coal,[231—234] and considerable ingenuity has been used in the design of equipment for use in well logging. A source of fast neutrons, generator or isotope, is shielded from one or more thermal neutron detectors and lowered down the borehole. Various devices have been developed for the correction of the reading for density,[235, 236] energy of the neutron source,[237] salinity,[238] and borehole diameter.[239] Apparatuses incorporating various time delays,[240, 241] partially overlapping time windows,[242] and difference techniques[243]

[231] Yu. V. Konoplev, O. A. Kitsenko, and N. V. Detkova, *Nuclear Sci. Abs.*, 1974, **30**, 6560.
[232] H. Sherman and J. Tittman, Canad. P. 925 227, 24 April 1973.
[233] E. Chruisciel, J. Massalski, K. Morstin, and A. Starzec, *Acta Geophys. Polon.*, 1973, **21**, 325.
[234] S. G. Komarov, 'Geophysical Methods for Investigating Boreholes', Izdatel'stvo Nedra, Moscow, 1973.
[235] J. R. Hearst, *Nuclear Sci. Abs.*, 1974, **30**, 6764, 6765.
[236] V. A. Emel'yanov and L. I. Beskin, *Tr. Vses. Nauch-Issled. Inst. Gidrotekh. Melior*, 1970, 32.
[237] V. S. Shlykov, *Nuclear Sci. Abs.*, 1973, **29**, 7727.
[238] V. E. Lebedev, L. G. Petrosyan, and D. M. Srebrodol'skii, *Nuclear Sci. Abs.*, 1974, **30**, 6559.
[239] N. A. Schuster, U.S. P. 3 784 822, 8 January 1974.
[240] M. P. Smith, U.S. P. 3 818 225, 18 June 1974.
[241] V. A. Yudin, L. B. Lutfirakhmanova, and V. Tseitlin, *Nuclear Sci. Abs.*, 1973, **29**, 7728.

have been used. The design of a neutron generator for pulsed neutron work[244] and a method of testing the quality of the neutron population data obtained from a well-log have been described.[245]

Neutron Activation Analysis. The principles and techniques of neutron activation analysis (NAA) are by now well established. The impact of the method, in industrial uses, has not been as great as might be expected, especially at the process end of the system. This is undoubtedly due to the fact that, until recently, the only useful source of neutrons was a nuclear reactor, and hence delays were involved in the sending and processing of samples for analysis. It has certainly been our experience (Petrochemicals Division of ICI) that the demand for NAA has increased by several orders of magnitude since the installation of a nuclear reactor on the plant site. The determination of elements in particular groups of samples, including samples of industrial interest, has been reviewed,[246] and several bibliographic publications on the applications of NAA are available.[247] Table 6[248—346] summarizes some of the

[242] W. W. Givens, U.S. P. 3 800 150, 26 March 1974.
[243] A. L. Polyanchenko, *Nuclear Sci. Abs.*, 1974, **30**, 6563.
[244] R. L. Caldwell, U.S. P. 3 800 145, 26 March 1974.
[245] L. A. Jacobson, W. B. Wall, and C. W. Johnstone, U.S. P. 3 868 505, 25 February 1975.
[246] M. Rakovic, 'Activation Analysis', Iliffe, London, 1970.
[247] N.B.S. Technical Note 467, 'Activation Analysis: A Bibliography through 1971', ed. G. J. Lutz, R. J. Boreni, R. S. Maddock, and J. Wing, U.S. Department of Commerce, August, 1972.
[248] H. A. Braler, *Analyt. Chem.*, 1975, **47**, 199R.
[249] A. R. Pouraghabagher and A. C. Profio, *Analyt. Chem.*, 1974, **46**, 1223.
[250] M. Tamura and T. Yamamoto, *Kogai Shigen Kenyusho Hokoku*, 1972, 1.
[251] R. Patek and R. Sorantin, *Allg. prakt. Chem.*, 1973, **24**, 87.
[252] R. Zaghoul, M. Obeid, and H. Staerk, *Radiochem. Radioanalyt. Letters*, 1973, **15**, 363.
[253] A. M. Passaglia, *Nuclear Sci. Abs.*, 1974, **29**, 28 968.
[254] A. M. Passaglia and F. W. Lima, *Rev. Brasil. Tecnol.*, 1973, **4**, 31.
[255] J. G. Larson and J. M. Orange, U.S.P. 3 723 732, March 27, 1973.
[256] H. Al-Shahristani and M. J. Al-Attiya, *Nuclear Sci. Abs.*, 1974, **29**, 4794.
[257] H. Al-Shahristani and M. J. Al-Attiya, *J. Radioanalyt. Chem.*, 1973, **14**, 401.
[258] I. D. Berkutova, E. S. Vuimova, T. Jalnina, I.M. Zlotova, and K. J. Yakubson, *J. Radioanalyt. Chem.*, 1973, **18**, 119.
[259] H. Vogg and H. Braun, *Nuclear Sci. Abs.*, 1974, **29**, 18 053.
[260] Y. Maki, T. Norjiri, and B. A. Masilungan, *Radioisotopes (Tokyo)*, 1974, **23**, 149.
[261] T. Vandlek, V. Klimint, and V. Scasnar, *Radioisotopy*, 1973, **14**, 537.
[262] K. Zietlow and K. Nagorny, *Kerntechnik*, 1973, **15**, 473.
[263] P. Martinelli, *Nuclear Sci. Abs.*, 1974, **29**, 10 334.
[264] C. Segebade, *Radiochem. Radioanalyt. Letters*, 1973, **15**, 260.
[265] *Amer. Ink Maker*, 1973, **51**, 31.
[266] G. G. Lutz, *Analyt. Chem.*, 1974, **46**, 618.
[267] A. Owlya, B. Parsa, and F. Fakhrvaezi, *Radiochem. Radioanalyt. Letters* 1974, **16**, 355.
[268] D. Fougea, M. Ghaleb, P. Gerald, and M. Pineri, *Rev. Gen. Caoutchouc, Plast.*, 1972, **49**, 1063.
[269] K. C. Campbell, J. V. Reid, and D. Gibbons, *Radiochem. Radioanalyt. Letters*, 1974, **16**, 283.
[270] J. W. Butler and R. H. Marsh, *Rubber Chem. Technol.*, 1972, **45**, 1560.
[271] R. F. Stewart, A. W. Hall, J. W. Martin, W. L. Farrior, and A. M. Poston, *Nuclear Sci. Abs.*, 1975, **31**, 16 647.
[272] D. R. Parsignault, H. H. Wilson, R. Mineski, and S. L. Blott, Neutron Sources Applications, Proc. American Nuclear Society Nat. Top. Meeting, 1971, Vol. 3, IV-40-IV-46.
[273] C. Block and R. Dams, *Analyt. Chim. Acta*, 1974, **71**, 53.
[274] L. Loska and L. Gorski, *Koks, Smola, Gaz.*, 1973, **18**, 52.
[275] C. Block and R. Dams, *Environ. Sci. Technol.*, 1975, **9**, 146.
[276] C. Block and R. Dams, *Analyt. Chim. Acta*, 1974, **68**, 11.
[277] J. Fahland, 4th Seminar on Activation Analysis of the Bundesanstalt für Materialprüfung, Berlin, October 1973.

[278] P. M. Santoliquido, *Radiochem. Radioanalyt. Letters*, 1973, **15**, 373.
[279] J. N. Weaver and D. J. von Lehmden, Symposium on Modern Methods of Fuel Analysis, University Park, Boston, Mass., U.S.A., 10 April 1972, p. 16.
[280] N. Tamura, *Radiochem. Radioanalyt. Letters*, 1974, **18**, 135.
[281] W. A. Straub and J. K. Hurwitz, *Analyt. Chem.*, 1975, **47**, 112R.
[282] H. J. Seim, R. C. Calkins, and J. A. Macksey, *Analyt. Chem.*, 1975, **47**, 128R.
[283] M. Burianova and J. Frana, *Radioisotopy*, 1973, **14**, 635.
[284] R. Van Grieken, *Verhandel. Koninkl. Vlaam. Acad. Wetenschap., Belg. Kl. Wetenschap.*, 1973, **35**, 1.
[285] S. H. S. Fulda, G. F. Palino, and A. V. Bellido, *Nuclear Sci. Abs.*, 1974, **30**, 23 502.
[286] I. Racataian, C. Postelnicu, G. Mociomita, L. Biciolla, and G. Brezeanu, *Nuclear Sci. Abs.*, 1975, **31**, 10 506.
[287] S. Dincer, H. Oezyol, H. Semimli, and E. Bantcugil, *Tech. J. Ankara Nuclear Res. Centre*, 1974, **1**, 111.
[288] M. Chiba and Y. Inoue, *J. Radioanalyt. Chem.*, 1974, **20**, 83.
[289] L. Alaerts, J. P. Op de Beeck, and J. Hoste, *Analyt. Chim. Acta*, 1973, **70**, 253.
[290] See ref. 165.
[291] C. Vandercasteele, R. van Grieken, and J. Hoste, *Analyt. Chim. Acta*, 1974, **72**, 31.
[292] V. A. Muminov, L. V. Navalikhin, and K. Khamrakulov, *Ind. Lab (U.S.S.R.)*, 1974, **40**, 69.
[293] S. Gangasharan, S. Yegnasubramanian, and S. C. Misra, *J. Radioanalyt. Chem.*, 1975, **24**, 57.
[294] B. Zitnansky, *Radioisotopy*, 1973, **14**, 525.
[295] V. I. Melent'ev, V. V. Ovechkin, and V. S. Rudenko, *Soviet J. Atomic Energy*, 1973, **34**, 45.
[296] K. Mukai, T. Takano, and K. Takada, ref. 290, p. 63.
[297] T. Mamuro and A. Mizhata, *Ann. Report Radiation Center Osaka Prefecture*, 1973, **13**, 23.
[298] A. A. Samadi, R. Grynszpan, and M. Fedorff, *Radiochem. Radioanalyt. Letters*, 1974, **19**, 171.
[299] L. Zapletal, F. Kukula, and V. Virish, *Nuclear Sci. Abs.*, 1974, **30**, 2440.
[300] T. S. Urbanski, *J. Radioanalyt. Chem.*, 1974, **23**, 13.
[301] D. Duffey, I. P. Balgna, P. F. Wiggins, and A. A. Elkardy, *Trans. Amer. Nuclear Soc.*, 1973, **17**, 120.
[302] G. Kolaski, E. Siewierski, and E. Vieczorek, *Isotopenpraxis*, 1973, **9**, 369.
[303] J. Csak, B. Vorsatz, L. Szabo, and E. Zemplen-Papp, *Banyasz. Kohasz. Lapok, Kohasz.*, 1973, **106**, 323.
[304] H. Zwittlinger, *J. Radioanalyt. Chem.*, 1973, **14**, 147.
[305] V. L. Shinkarchuk, L. V. Navalikhin, N. Yu. Talanin, and M. Allaniyazov, *Izotopy S.S.S.R.*, 1972, 23.
[306] B. F. Schmitt (comp.), Symposium on Nuclear Analytical Techniques in the Production and Industrial Use of Noble Metals, Brussels, November 1971.
[307] K. W. Dolan and L. W. Dahlke, *Nuclear Sci. Abs.*, 1975, **31**, 8584.
[308] T. B. Pierce, ref. 290, p. 49.
[309] R. Zaghloul, M. Obeid, and R. Henkelmann, *Radiochem. Radioanalyt. Letters*, 1974, **17**, 7.
[310] C. M. Lobanov, Yu. A. Levushkii, and S. P. Vlasyuga, *Doklady Akad. Nauk Belorussk. S.S.R.*, 1973, **17**, 905.
[311] G. M. Sanquist, O. Akalin, V. C. Rogers, and T. A. Linn, jun., *Trans. Amer. Nuclear Soc.*, 1974, **18**, 65.
[312] L. M. Velyus, V. T. Fomenko, and I. N. Fomenko, *Ind. Lab. (U.S.S.R.)*, 1972, **38**, 1882.
[313] D. G. Campbell, P. D. Farm, and C. V. Gladysz, *C.I.M. Bull.*, 1974, **67**, 90.
[314] E. M. Radisov, V. V. Miller, and Yu. S. Shimilevich, *Nuclear Sci. Abs.*, 1975, **30**, 6807.
[315] V. A. Volokh, *Nuclear Sci. Abs.*, 1974, **29**, 7505.
[316] I. Ahmad and D. F. C. Morris, *Radiochem. Radioanalyt. Letters*, 1974, **16**, 303.
[317] R. Gijbels, *Mineral. Sci. Eng.*, 1973, **5**, 304.
[318] A. N. Gorbachev, A. M. Karpunin, and L. F. Matukanis, *Nuclear Sci. Abs.*, 1973, **29**, 29 005.
[319] A. A. El-Kady and I. Hamouda, *Nuclear Sci. Abs.*, 1974, **30**, 14 741.
[320] V. Moucka and J. Kucera, *Nuclear Sci. Abs.*, 1974, **29**, 15 231.
[321] P. A. Vagonov, V. S. Ponomarev, A. V. Eakhtiarov, V. A. Meier, A. N. Zhukovskii, and R. I. Plotnikov, *Uch. Zap. Leningrad. Gosud. Univ., Ser. Fiz. Geol. Nauk*, 1973, 223.
[322] I. M. Cohen, *Radiochem. Radioanalyt. Letters*, 1973, **15**, 379.
[323] E. Steinnes, *Analyt. Chim. Acta*, 1974, **68**, 25.
[324] E. E. Rakovskii, T. D. Krylova, and G. M. Baesskaya, *Zhur. analit. Khim.*, 1974, **29**, 2166.
[325] E. G. Agudo, V. Duarte, and C. L. Seignemartin, *Nuclear Sci. Abs.*, 1974, **30**, 31 471.
[326] E. Garcia Agudo, C. L. Seignemartin, and U. Duarte, *Nuclear Sci. Abs.*, 1974, **30**, 28 913.
[327] N. Das and S. N. Bhattacharyya, *Talanta*, 1974, **21**, 894.
[328] G. Santos, jun. and M. Yulo, *Nuclear Sci. Abs.*, 1975, **31**, 10 504.
[329] V. V. Gorshkov, N. N. Basargin, L. A. Smakhtin, and T. S. Sinitsyna, *Ind. Lab. (U.S.S.R.)*, 1974, **40**, 226.

Table 6 *Neutron activation analysis of industrial materials*

Elements determined	Neutron source	Sample	Ref.
A. Oil and petroleum products[248]			
S	^{252}Cf	Oil (0.5% S upwards)	249
O, S, N	Generator	Petroleum hydrocarbons	250
V, I, Cl, Br, Mn, Na, Au, Cr, Fe, Sc, Ag, Zr, Se, Sb, Co	Reactor	Austrian crude oil	251
V, S, Na, Br	Reactor	Arabian crude oil	252
V	Reactor	Petroleum	253—255
Al, V, Cl, Ni, Mn, K, Zn, Ga, Na, As, Br, Hg, Co	Reactor	Iraqi oil	256
Al, A, As, Au, Br, Ca, Cl, Co, Cr, Cu, Eu, Fe, Ga, Hg, K, La, Mn, Mo, Na, Ni, Re, S, Sb, Sc, Se, Sm, V, W, Zn	Reactor	Iraqi crude oil	257
Various elements	Reactor	Petroleum	258
B. Chemical products etc.			
F	^{252}Cf	Teflon	259
F	^{241}Am/Be	Plastics	260
O, P, Ba	Generator	Oil additives	261
O	Reactor	Polyethylene	262
Cl	Isotope	Fibres	263
Br	Generator	Pesticides	264
As, Cd, Hg, Pb, Se, Ba	Reactor	Pigments	265
Pb	^{252}Cf	Paint	266
Cl, Mn, Na, K, Br, La, Au, Cr, Fe, Cs, Sc, Zn, Co	Reactor	Margarine	267
O	Reactor	Rubbers	268
Cl	Reactor	Rhodium catalysts	269
Cl	Reactor	Rubber	270
V	Reactor	Cracking catalysts	253, 254
C. Coal and coal products			
S	^{252}Cf	Coal streams	271, 272
Si, O	Generator	Coal, coal ash	273

[330] M. A. Belyakov, Eh. P. Terent'ev, and V. A. Akındınov, *Khim. Prom.*, 1974, **1**, 44.
[331] IAEA Report R-1137-F (*Nuclear Sci. Abs.*, 1975, **31**, 10 504).
[332] E. Stiennes and A. D. Mukherjee, *J. Radioanalyt. Chem.*, 1973, **14**, 129.
[333] S. E. Chao, W. D. Lu, and S. C. Wu, *Nuclear Sci. J. (Taiwan)*, 1974, **11**, 137.
[334] M. J. Fishman and D. E. Erdmann, *Analyt. Chem.*, 1975, **47**, 334R.
[335] H. A. van der Sloot and H. A. Das, *Analyt. Chim. Acta*, 1974, **73**, 235.
[336] J. Schneider and R. Geisler, Trace Analysis, Gesellschaft Deutscher Chemika, Erlangen, Germany, 2 April 1973.
[337] D. M. Bibby and G. Oldham, *J. Radioanalyt. Chem.*, 1974, **20**, 101.
[338] S. Specht, in ref. 277.
[339] J. L. Verot and J. J. Jaumier, *J. Radioanalyt. Chem.*, 1973, **17**, 237.
[340] E. Bruninx, *Analyt. Chim. Acta*, 1973, **67**, 17.
[341] W. Zmijewska, *Chem. analit. (Warsaw)*, 1973, **18**, 817.
[342] J. T. Tanner, H. Melvin, and G. E. Holloway, *Analyt. Chim. Acta*, 1973, **66**, 456.
[343] C. Gountchev, A. Fourcy, M. Berchard, R. Bittel, and D. Schaal, *Nuclear Sci. Abs.*, 1974, **29**, 9727.
[344] P. Ostermann, H. Staerk, and R. Henkelmann, in ref. 277.
[345] T. T. Vandergraaf and A. G. Wikjord, *Nuclear Sci. Abs.*, 1974, **29**, 15 205.
[346] P. Cukor, *Analyt. Chem.*, 1976, **48**, 51A.

Table 6 *continued*

Elements determined	Neutron Source	Sample	Ref
Si, Al	Generator	Hard coal	274
44 elements	Reactor, Generator	Belgian coal, coal ash	275, 276
Si, O	—	Lignite	277
As	Reactor	Coal ash	278
Hg	Reactor	Coal	279
Various elements	—	Coal	280
D. Metals and metallurgical products[281, 282]			
Rare earths	Reactor	Steel	283
Si, Cu, P, O	Generator	Iron, steel	284
O	Generator	Steel	285—287
Si	Generator	Cast iron	288
Si, Al	^{277}Ac/Be, ^{241}Am/Be	Ferrasilicon, iron ore	289, 290
Cr, Si	Generator, ^{241}Am/Be	Steel, chrome concentrates	290, 291
Si, Cr, Fe, Mn	Generator	Steel, alloys	292
F, Al, Si, P, K, Mn, Mo, W	Generator	Steel	293
O, Fe, Mg, Mn	Generator	Fe alloys, welding fluxes	294
O, F	Generator, ^{238}Pu/Be	Zr, Ta, Al	295, 296
Al, V, Cu, Co, Mn	Reactor	Fe	
Al, Sc, Cr, Fe, Co, Zn, As, Sb, Ba, La, Ce, Sm, Tb, Lu, Hf, Th			297
Cu, Mn, Pd, Pt, Ir, Ru, Os	Reactor	Al alloys	298
La	Reactor	Steel, slag	299, 300
Cu	^{252}Cf		301
La, Sm, Sc, Lu	Reactor	Steel, slag	302
O	Reactor	Al, alloys	296, 303
Cr, Mn, Fe, Co, Ni, W	Reactor	Refined steel	304
W, Co, Ti	Generator	Alloys	305
Various	—	Noble metals	306
Cu, Ag,	^{252}Cf	Cusil braze	307
V	^{252}Cf	Steel	308
E. Ores and mineral products			
Fe	^{241}Am/Be	Iron ore	309
Sc, Cr, Fe, Co, Zn, Rb, Sr, Zr, Sb, Cs, Ba, La, Ce, Nd, Sm, Eu, Gd, Tb, Tm, Yb, Lu, Hf, Ta, Th	Reactor	Various ores and minerals	310
W	^{252}Cf	Scheelite	259
Cu	Isotope source	Cu process streams	311, 312
Si	—	Iron ore	313
Various	Po/Be	Sand, gypsum	314
Fe, Al	Po/Be	Phosphonites	315
Ag, Rh, Ir, Au	^{241}Am/Be	Precious-metal concentrates	316
Various	Isotope sources, generators	Ores, minerals, rocks	317
F, Sb, Hg, Sn, Be, W, Mo	Po/Be	Fluorite, apatite, phosphorite	318
Be, Ag, Au, W, Sc, Ce, Pb, Ta, N, Sn	Reactor, ^{238}Pu/Be	Various ores	319
U, Th	Reactor	Various minerals	320
Ta, Eu, Ce, Hf, Tl	Reactor	Ores	321
In, Cd, Mn	Reactor	Blendes	322

Table 6 *continued*

Elements determined	Neutron Source	Sample	Ref
Ni	Reactor	Various standard rocks	323
Noble metals	Reactor	Natural and commercial products	324
Ta	Reactor	Pegmatite ores	325
Au	Reactor	Various minerals	326—328
Ag	Reactor	Ores	329
P	—	Raw materials	330
V, Mn	Reactor	Magnetite sands	328
U	Reactor	Co/Mo ores	331
Sb, As	Reactor	Cu ores	331
Sc, Cr, Co, Zn, As, Se, Sb, Ir, Au, Th	Reactor	Pyritic lead–zinc ores	332
U, Th, rare earths	Reactor	Yellow monazite	333
F. Water[334]			
Hg	Reactor	—	335
As, Cd, Cu, Hg, Sb, Zn	Generator, ^{252}Cf Reactor	Drinking water	336
Na, Al, P, Cr	Generator	—	337
Heavy metals	^{252}Cf	Waste waters	338
G. Miscellaneous			
As, Se, Te	Reactor	Sulphur	339
Ga, Se	Reactor	Semiconductors	340
Ag, Cr, In, Fe	Reactor	Silicon semiconductor	341
As, Sb	Reactor	Laundry oils	342
Br, As, Hg, F	Reactor	Food	343
B	Reactor	Glass	344
Various elements	Reactor	Silicon carbide	345
Si, Al	^{241}Am/Be	Cement	290
Si, Al	^{241}Am/Be	Paper pulp	290
Si, O	Reactor	Resists made from silicones	346
O, Cl	Reactor	Plastic wire coating	346

more recent publications on samples of industrial importance. It will be noted that a number of the references in Table 6 use neutron sources other than nuclear reactors. It is these applications which are suited to on-line measurements.[347] The principles and applications of isotopic sources of neutrons have been reviewed for ^{252}Cf,[308, 348—354] α-emitters–light element combinations,[355—357] and neutron generators.[308, 358—360] The advent of californium-252 sources could well revolu-

[347] J. Perdijon, *Talanta*, 1974, **21**, 1047.
[348] H. Braun and F. Riffl, in ref. 277 (*Nuclear Sci. Abs.*, 1974, **29**, 18 554).
[349] J. L. Vervt, *J. Radioanalyt. Chem.*, 1974, **19**, 177.
[350] *Nuclear Sci. Abs.*, 1974, **30**, 26 450.
[351] *Californium-252 Progress*, Nos. 1—18.
[352] W. R. Cornman, W. C. Reinig, and P. H. Permar, *Nuclear Sci. Abs.*, 1974, **29**, 12 991.
[353] J. John, *Trans. Amer. Nuclear Soc.*, 1974, **18**, 91.
[354] G. J. Lutz, *Trans. Amer. Nuclear Soc.*, 1973, **17**, 125.
[355] L. Alaerts, J. P. op de Beeck, and J. Hoste, *Analyt. Chim. Acta*, 1974, **69**, 1.
[356] P. B. Pawaskar, G. R. Reddy, and M. Sankar Das, *Nuclear Sci. Abs.*, 1975, **31**, 25 198.
[357] I. Josza, G. Peto, G. Csikai, and K. D. Vinhlong, *Kozlem*, 1974, **16**, 339.
[358] D. E. Wood, *Trans. Amer. Nuclear Soc.*, 1974, **18**, 746.
[359] S. S. Nargolwalla and E. P. Przybylowicz, 'Chemical Analysis', Vol. 39, John Wiley, New York, 1973.
[360] M. Brafman, A. Godeau, J. Laverlochere, and J. L. Lecote, Fr.P. 2 168 238, 21 January 1972.

tionize the applications of on-line NAA. Neutron generators have made similar strides in their possible applications, and sources producing neutron fluxes comparable with these from nuclear reactors may now be prepared, with guaranteed lifetimes of 1000 h. The optimization of NAA systems under various conditions of continuous flow, where one or more elements may be determined,[361, 362] and under cyclic conditions[363] has been discussed. All NAA analyses consist of three basic operational units; the activation unit (source + cell), the detector, and the electronics unit. Provision is usually made for the thermalization of the neutron source, unless fast neutron activation is required. Several possible combinations of electronics units are possible.[308]

Well Logging.—Radioisotopic methods have been extensively used for carrying out various measurements in boreholes, storage wells, and cavities. The techniques applied have all been previously described, but special equipment and measuring techniques need to be devised because of the adverse conditions under which the instruments must operate, such as high pressures and temperatures. The need to lower the equipment down relatively narrow casings has led to the development of special miniature versions of more conventional equipment.

Measurements carried out vary from the relatively simple interface or level detection using γ-backscatter to the more complex techniques of geological logging using γ-backscatter,[364, 365] the determination of density,[366—370] for locating ore bodies,[371, 372] and for determining composition.[373, 374]

Inelastic neutron scattering and detection of the resultant γ-rays has been used to measure C/O and Ca/Si ratios in order to identify oil zones and oil saturation under a variety of conditions.[375—378] A similar apparatus has been applied to the location of coal and oil-shale zones.[379] The porosities of clays,[380] carbonates,[381] and other

---

[361] L. Laska, J. Janczyszyn, and L. Gorski, *J. Radioanalyt. Chem.*, 1973, **14**, 11.
[362] UKAEA, B.P. 1 321 869, 4 July 1973.
[363] Y. Maki, T. Norjiri, and B. A. Masilungan, *Radioisotopes* (*Tokyo*), 1974, **23**, 149, 870, 1321.
[364] D. F. Gera, N. I. Moroz, L. S. Shastova, and V. A. Kraznoperov, *Razved ka i Okhr and Nedr*, 1973, 33.
[365] I. P. Borzyak, *Izvest. V. U. Z., Geol. Razved*, 1974, 119.
[366] S. I. Vasil'ev, *Nuclear Sci. Abs.*, 1974, **29**, 7730.
[367] A. Heslop, Fifteenth Annual Logging Symposium Transactions, Houston, Texas, Soc. of Professional Well-log Analysts, 1974, p. M1.
[368] I. G. Dyad'kin, B. N. Krasil'nikov, and N. Starikov, *Soviet Atomic Energy*, 1973, **35**, 936.
[369] N. Ya. Basin, L. I. Bermann, P. T. Parfenov, and P. T. Savinkin, *Razved. Geofiz.*, 1973, **55**, 167.
[370] K. Preiss and A. Lahat, *Geotechnique*, 1972, **22**, 663.
[371] B. N. Krasil'nikov, *Nuclear Sci. Abs.*, 1974, **29**, 7732.
[372] P. M. Vol'fshtein, Yu. P. Yanhevski, V. V. Belyi, A. P. Ochkur, E. V. Egorov, and E. A. Sokolov, *Nuclear Sci. Abs.*, 1974, **30**, 316.
[373] V. M. Moroshnichenko, *Razved. Geofiz.*, 1973, 170.
[374] J. Tittman, U.S. P. 3 864 569, 4 February 1975.
[375] H. D. Smith and W. E. Schultz, ref. 367, p. K1.
[376] W. E. Schultz and H. D. Smith, *J. Petrol. Technol.*, 1974, **26**, 103.
[377] R. W. Pitts, U.S. P. 3 842 265, 15 October 1974.
[378] W. E. Schultz and H. D. Smith, U.S. P. 3 838 279, 24 September 1974.
[379] P. F. McKinlay, U.S. P. 3 849 646, 19 November 1974.
[380] Yu. P. Bal'vas, Ya. N. Basin, N. K. Kukharenko, and Yu. V. Tyukaev, *Razved. Geofiz.*, 1973, 109.
[381] T. G. Kerimov and P. T. Kotov, *Razved. Geofiz.*, 1973, 161.

rocks[382, 383] have also been measured by this method. The determination of the ratio of Al to Si in large mass alumino-silicate rocks has been carried out, using a 14 MeV neutron source, which shows a number of advantages over other methods[384] such as the two-source method. The method has also been used for the location of gas deposits[385] and the detection of gas/water and oil/water interfaces.[369] Neutron activation analysis has been used for the determination of aluminium and iron in bauxites,[386] fluorine in phosphate deposits,[387] iron and titanium in bauxites,[388, 389] and of silicon[390, 391] and rare metals[392] in various rocks. Several patents cover apparatus for carrying out NAA well logs.[393, 394]

Several methods for moisture and hydrogen determinations as applied to well logging have already been mentioned in the sections on neutron thermalization techniques. Neutron lifetime measurements[395—397] and neutron-capture[398—400] techniques have also been used.

Direct measurement of the natural γ-activity of rock strata has been used for the measurement of the clay content of silt-sandstone[401] and the clay in Devonian carbonate rocks.[402]

The whole field of well logging[403] and nuclear techniques[404] for the evaluation of mineral deposits has been reviewed.

Radiography.—γ-*Radiography.* It would be impossible within the scope of this article to review adequately the many applications of industrial radiography. Suffice it to say that, despite being a long-established technique, developments both in source technology[405—407] and in image-processing methods[408] are continuing.

---

[382] R. G. Beil, U.S. P. 3 833 809, 3 September 1974.
[383] V. M. Lakhmyuk, *Dopovidi Akad. Nauk Ukrain R.S.R.*, Ser. B, 1974, 704.
[384] V. K. Andreev, A. A. Barenbaum, B. G. Egiazarov, Yu. P. Sel'dyakov, and K. I. Yakubson, 'Yadernoe Priborostroenie', Atomizdat, Moscow, 1972, p. 95.
[385] K. A. Grudkin, *Nuclear Sci. Abs.*, 1974, 30, 6561.
[386] V. I. Ishchenko and A. V. Leikin, *Nuclear Sci. Abs.*, 1974, 30, 322.
[387] R. D. Belenki, *Nuclear Sci. Abs.*, 1974, 30, 324.
[388] O. V. Shishakin, I. P. Koshelev, A. P. Taushkanov, G. I. Shepelev, and G. P. Katorcha, *Nuclear Sci. Abs.*, 1974, 30, 323.
[389] O. V. Shishakin and A. P. Taushkanov, 'Problems of Ore Geophysics in Korzakhistan., Alma-Ata, Kazakhskij Filial, Voesayuzrujj Inst. Razved. Geofiziki, 1973, p. 279.
[390] O. V. Shishakin, A. P. Taushkanov, and A. A. Teben'kov, *Nuclear Sci. Abs.*, 1975, 31, 11 271.
[391] H. J. Paap and H. L. Tanner, U.S.P. 3 781 545, 25 December 1973.
[392] A. M. Kinehenko, F. L. Nastich, and Z. G. Muromtsev, *Nuclear Sci. Abs.*, 1974, 30, 325.
[393] H. E. Hall, jun., A. S. McKay, and H. J. Paap, Danish P. 989/72, 4 December 1972.
[394] D. M. Arnold and R. W. Pitts, jun., U.S. P. 3 842 264, 15 October 1974.
[395] V. N. Moiseev and E. V. Varvarskii, *Nuclear Sci. Abs.*, 1974, 30, 6562.
[396] E. C. Hopkinson, A. H. Youmans, and R. B. Johnson, in ref. 367, p. AA1.
[397] E. Chrisciel, T. Dabek, J. Massalski, K. Morstin, T. Owsiak, and A. Starzec, *Nuclear Sci. Abs.*, 1974, 29, 482.
[398] A. A. Brem, E. I. Krapivskii, and V. B. Sal'tsevich, *Express-Inform. Ser., Razved. Prom. Geofiz.*, 1973, 1.
[399] J. D. Robinson, U.S. P. 3 852 593, 3 December 1974.
[400] D. M. Arnold and R. W. Pitts, jun., Canad. P. 951 438, 16 July 1974.
[401] V. D. Sharabarin and M. S. Sokolov, *Izvest. V.U.Z., Neft. Gaz.*, 1973, 9.
[402] A. P. Anpilogov, *Doklady Akad. Nauk Belorussk. S.S.R.*, 1973, 17, 952.
[403] J. H. Scott and B. L. Tibbetts, *Nuclear Sci. Abs.*, 1975, 31, 16 545.
[404] A. M. Blyumentsev, J. I. Fel'dman, K. I. Yakubson, and I. N. Senko-Bulatniji, *Nuclear Sci. Abs.*, 1974, 30, 6557.
[405] Y. Sumiya and M. Anma, *Nuclear Sci. Abs.*, 1974, 30, 18 555.
[406] T. Tsujimoto, T. Yoshimoto, and K. Katsurayama, *Nuclear Sci. Abs.*, 1974, 30, 18 550.

The field of γ-radiography has been reviewed.[409, 410]

Neutron Radiography. Neutron radiography, a technique complementary to γ-radiography, has proved especially valuable for the detection of light elements. The technique has been recently reviewed.[411-414]

The choice of neutron source and energy for different applications has been discussed,[413, 415] and the use of a 35 MeV linear electron accelerator for the production of thermal and epithermal neutron fluxes outlined.[416] Various imaging devices have been used in neutron radiography, ranging from film techniques[417] to scintillation screens such as gadolinium oxysulphide[418, 419] or ^{6}LiF–ZnS(Ag) or Co tungstate to video-electronic image analysis of the resulting image.[420]

The applications of on-line and portable neutron radiography systems have been outlined,[421] and particular applications to the inspection of nuclear reactor fuel assemblies[422-424] and the measurement of cadmium in single crystals of $Hg_{1-x}Cd_x Te$[425] and of boron in steels have been described.[426]

[407] V. Vondruska, Institute for Research, Production, and Uses of Radioisotopes, Prague, October 1971 (*Nuclear Sci. Abs.*, 1974, **29**, 2608).
[408] J. A. Holloway, W. L. Shelton, and J. P. Mitchell, Sixth National SAMPE Technical Conference, Vol. 6, Azusa, CA; Soc. for Advancement of Material and Process Engineering 1974, p. 110.
[409] I. Panaitesui, *Nuclear Sci. Abs.*, 1975, **31**, 5928.
[410] 'Handbook of Radiographic Apparatus and Techniques', 2nd edn., Cambridge Welding Institute, 1973.
[411] N. van der Klej and H. P. Leeflang, *Materialprüfung*, 1974, **16**, 287.
[412] J. J. Rozental, 3rd Fabrication Inspection Meeting, Rio de Janeiro, October 1973 (*Nuclear Sci. Abs.*, 1975, **31**, 11 202).
[413] J. Rant, M. Copic, V. Dimic, R. Ilic, and F. Sirca, 2nd European Conference of Triga Reactor Users, Pavia, Italy, September 1972.
[414] Y. D. Dande, *Nuclear Sci. Abs.*, 1975, **31**, 16 482.
[415] Z. Hrdlicka, *Nuclear Sci. Abs.*, 1974, **29**, 2607.
[416] J. S. Hewitt, M. K. Aydogdu, G. R. Blumenauer, and H. A. Robitaille, *Non-Destructive Testing*, 1974, **7**, 315.
[417] H. Schuelken, *Kerntechnik*, 1974, **16**, 365.
[418] K. L. Swinth and L. L. Nichols, *Nuclear Sci. Abs.*, 1974, **29**, 321.
[419] K. L. Swinth, *Nuclear Sci. Abs.*, 1974, **29**, 24 133.
[420] A. Vary and K. J. Bowles, *Material Eval.*, 1974, 7.
[421] S. Sciutti, 18th Nuclear Congress, Rome, March 1973.
[422] N. P. Lapinski and H. Berger, *Trans. Amer. Nuclear Soc.*, 1973, **17**, 178.
[423] W. J. Stamn and N. van der Kley, 'NDT for Reactor Core Components and Pressure Vessels', Vienna, Austria, November 1971, p. 87.
[424] R. W. McChing, in ref. 423, p. 155.
[425] G. Gasparrini, M. Mangialoyo, D. Passoni, and B. Pirovano, *Infrared Phys.*, 1974, **14**, 145.
[426] L. Boutaine, *Mem. Sci. Rev. Met.*, 1973, **70**, 747.

2
Activation Analysis in Archaeology

BY G. HARBOTTLE

1 Introduction

We may be fairly certain that the first chemist to carry out an analysis of any artifact of archaeological origin was Martin Klaproth (1743—1817). Before the eighteenth century had ended, he had reported detailed analyses of three Roman glass tesserae, coloured red, green, and blue, from a mosaic in the Villa of Tiberius at Capri.[1] It is interesting that both in his work, and in later studies by Sir Humphrey Davy,[2] the purpose was to determine the cause of the colour. Thus originated one of the central motives of archaeological chemical analysis, which has persisted to the present day, namely, the understanding of ancient technology.[1,3,4]

A second motive may be described broadly as 'provenience research', the identification of sources of objects by the observation of similarities in chemical composition. In the words of E. T. Hall, 'Studies of trade routes both locally and between different countries can be helped by trace-element analysis. If we can trace a certain type of objects by its chemical composition, the fact that from stylistic considerations such objects were made in several places becomes unimportant—we may still be able to say whether different objects were made in the same place or not'.[5] And, one may add, in fortunate cases we may be able to trace objects to their actual origin, a clay bed, either isolated or in a river system, an ore deposit, obsidian outcrop, *etc*.

We do not surely know who first conceived the idea of provenience identification *via* chemical analysis, with its tacit assumption that materials of the same type but from different sources would be chemically different. A few years ago, with considerable excitement, I encountered the following passage in 'Murray's Guide to Greece'[6] under the entry for 'Thera' (the present-day Santorin) 'An increase in the demand for pozzolana . . . led to the discovery, 100 ft. below the surface, of the ancient settlement called by M. Fouqué a "Prehistoric Pompeii". The following is a summary of his article in the "Spectator" for Nov. 6, 1869 . . .'. There follows a description of Fouqué's discoveries, ending with the phrase 'It has been proved by a *chemical analysis* of the clay that the greater part of the pottery must have been manufactured in Thera itself'. Alas for the reputation for accuracy of the Murray's Guides; a copy of the 'Spectator' article yielded only the following: 'Their pottery must have come

---

[1] E. R. Caley, 'Analysis of Ancient Glasses 1790—1957', The Corning Museum of Glass, Corning, New York, 1962, p. 13.
[2] H. Davy, *Phil. Trans.*, 1815, **105**, 97.
[3] A. Lucas, 'Ancient Egyptian Materials and Industries', 4th Edn., revised by J. R. Harris, Arnold, London, 1962.
[4] E. R. Caley, 'Analysis of Ancient Metals', Pergamon Press, Oxford, 1964.
[5] E. T. Hall, *Phil. Trans.*, 1970, **A269**, 135.
[6] J. Murray, 'Murray's Guide to Greece', London, 7th Edn., 1900, p. 928.

from abroad, as the clay for its manufacture was not to be found in the island'. The final word, of course, is that given by Fouqué himself, 'La plupart de ces vases n'ont pas été fabriqués sur le sol où ils ont été retrouvés; ils ont été apportés du dehors, car à Santorin pas plus qu'à Therasia on n'aperçoit aucune substance argileuse. La cendre volcanique qui y abonde n'est point plastique, et la composition chimique en est d'ailleurs assez différente de celle de la matière des vases. Il faut nécessairement admettre que les poteries trouvées si abondamment sous le tuf ponceux de Santorin et de Therasia provenaient en grande partie de l'extérieur.'[7] This certainly contains the germ of the concept of provenience attribution (or its converse) *via* chemical composition, and Fouqué may well be given the credit, after all.[8]

At the end of the 19th century, we have the report of Richards at Harvard:[9] 'At the request of Mr. Edward Robinson, of the Boston Museum of Fine Arts, several analyses of ancient Athenian pottery were recently made at this laboratory. . . . the interest of these analyses was mainly archaeological, turning upon the identity of the source of these remains with that of others found in other cities, . . .' and finally 'The variations in the relative amounts are singularly small, the range being not nearly so large as that given by Brougniart, in his "Traité des Arts Ceramiques". Hence, it is possible, that all of these specimens, which were picked up in the city of Athens itself, were the product of a local pottery.'

There are several other reasons why chemical analyses of archaeological objects are undertaken. Two are mentioned by Caley, in 'Analysis of Ancient Metals', and of these the first involves coins specifically. 'Because coins were the only form of money used in ancient civilizations, the relationship between the coinage and economic conditions was close. Changes in the fineness and weight of ancient coins are therefore indicative of changes in these conditions. For example, a progressive decrease in the fineness, weight or both of the silver coins of a given series indicates the occurrence of monetary inflation and a rise in prices during the period when the decrease occurred.'[10] He gives an example drawn from the Roman Empire; the cynic might add to this the present-day substitution of copper-nickel 'sandwich' coins for silver in the U.S.A.

Finally, there is the obvious need for chemical analysis in the detection of forgeries, although this might be considered a form of provenience investigation. Caley's book on ancient metals contains excellent bibliographies of the literature referring to all of the above-listed types of archaeological analysis, arranged according to the nature of the metals (gold, silver, copper, non-ferrous, iron, steel *etc*.) analysed, and it is interesting to see numerous references to experiments in the mid-nineteenth century and even earlier.

The nuclear age was not far advanced when von Hevesy and Levi, working in Copenhagen, published two papers[11] launching both the fields of neutron activation analysis, which today is in routine use in hundreds of laboratories, producing more than a thousand scientific reports a year,[12] and analysis *via* neutron absorption,

[7] M. Fouqué, 'Revue des Deux Mondes', Paris, 1869, **83**, 923.
[8] H. W. Catling, *Phil. Trans.*, 1970, **A269**, 175.
[9] T. W. Richards, *J. Amer. Chem. Soc.*, 1895, **17**, 152.
[10] See ref. 4, p. 163.
[11] G. Hevesy and H. Levi, *Kgl. Dansk Vidensk. Selskab., Mat.-Fys. Medd.*, 1936, **14**, 3; *ibid.*, 1938, **15**, 14.
[12] 'Activation Analysis: A Bibliography', ed. G. L. Lutz, R. J. Boreni, R. S. Maddock, and W. W. Meinke, N.B.S. Technical Note 467, National Bureau of Standards, Washington, 1968.

Activation Analysis in Archaeology

which has since lain almost untouched. At nearly the same time Seaborg and Livingood demonstrated that bombardment with charged particles, as well as with neutrons, could be a powerful analytical tool.[13]

Had it not been for two purely technical developments, NAA, at least, would have remained a laboratory curiosity. The first of these was the war-time development of the nuclear reactor, pushing the neutron fluxes available to the scientist up by orders of magnitude from the puny levels previously developed with radium–beryllium or, better, with moderated cyclotron neutrons, to the 10^{13} cm^{-2} s^{-1} in routine use today. This was important inasmuch as the sensitivity of detection is directly related to the flux level. A second development was the sodium iodide, and later germanium, gamma detector, with its attendant pulse-height analyser system, which allowed the discrimination and measurement of radioisotopes directly through their characteristic gamma rays, eliminating the need in many cases to carry out laborious chemical separations, and tediously to follow decay-curves as a means of identification. The sodium iodide counter was also the first *efficient* gamma detector, a point which is often forgotten.

The first papers reporting the application of nuclear techniques to archaeological materials were those of Ambrosino *et al.*[14, 15] and Emoto,[16] dealing with coins, the detection of silver in an ancient lead roof tile, and in the gold foil attached to a Korean sword. Other early work is cited in the bibliography of Sayre and Meyers.[17] The Chemistry Department of Brookhaven National Laboratory pioneered the systematic use of nuclear activation in archaeological provenience studies: it happened as follows. In the autumn of 1954 J. R. Oppenheimer, Director of the Institute for Advanced Study at Princeton, telephoned R. W. Dodson, the Chairman of the Chemistry Department at Brookhaven, suggesting the possible use of trace element analysis *via* neutron activation in establishing the provenience of archaeological ceramics, a problem Oppenheimer had discussed with archaeologists. Dodson visited Princeton shortly thereafter to discuss the project in more detail, and in the spring of 1955 enlisted E. V. Sayre to undertake experimental work. Sherds of pottery from the Mediterranean region (Asia Minor, Greece, and Italy) were made available by Professors H. A. Thompson of the Institute and H. Comfort of Haverford College, and Sayre analysed these for sodium and manganese, using a sodium iodide detector coupled to a 100-channel analyser. The results, which were expressed as ratios of ^{56}Mn:^{24}Na, showed distinct differences between sherds from different sources, but similarities between sherds from the same region. These were reported by Sayre and Dodson to a group made up of archaeologists, chemists, and Professor Oppenheimer at the Institute in March, 1956, and eventually published.[18] The results were considered so promising that additional research was encouraged by the group attending the meeting. By 1958 Sayre, Murrenhoff, and Weick had extended the method to Fine Orange ware from the Mayan civilization, and much additional material from the Roman world.[19] Elements measured also included lan-

---

[13] G. T. Seaborg and J. J. Livingood, *J. Amer. Chem. Soc.*, 1938, **60**, 1784.
[14] G. Ambrosino and P. Pindrus, *Rev. Metallurgy*, 1953, **50**, 136.
[15] G. Ambrosino and A. P. Weill, *Bull. Lab. Musée du Louvre*, 1956, **1**, 53.
[16] Y. Emoto, *Sci. Papers Japan, Antiques*, 1957, **13**, 37.
[17] E. V. Sayre and P. Meyers, *Art Archaeol. Technical Abs.*, 1971, **8**, 115.
[18] E. V. Sayre and R. W. Dodson, *Amer. J. Archaeol.*, 1957, **61**, 35.
[19] E. V. Sayre, A. Murrenhoff, and C. F. Weick, BNL Report 508, 1958; 'Proceedings of the Boston Museum of Fine Arts Seminar, September, 1958', pp. 153—180.

thanum, scandium, and chromium; element profiles were plotted for matching, and the 'best relative fit' technique was devised. At this time the Oxford group also began to analyse archaeological materials, by emission spectroscopy[20] as well as nuclear activation.[21] Meanwhile, Sayre's interests turned to the study of ancient glass.[22,23]

The next great advance in technique came with the discovery of the germanium detector, with its (to a nuclear scientist) astonishing increase in gamma-ray resolution over sodium iodide. This detector put activation analysis 'on the map': Sayre's paper entitled 'Refinement in Methods of Neutron Activation Analysis of Ancient Glass Objects through the Use of Lithium Drifted Germanium Diode Counters' was in fact the first to report the use and advantages of the 'Ge–Li' detector in activation analysis. Since then, the speed and power of the neutron activation–Ge–Li counting combination have given it a profound impact on analytical (especially trace-level) studies in biology, medicine, agriculture, the environment, geochemistry, lunar science, forensics, metallurgy, and, the subject of this essay, archaeology. The intended audience for this Report is, on the one hand, archaeologists, to show them what can be accomplished, and on the other, nuclear or physical scientists, to assist them in a practical way if they hope to begin such studies.

To this end, I will first discuss nuclear activation, making specific reference to problems encountered in archaeological studies, then tackle the crucial question of data-handling—interpreting the mass of analytical results generated in provenience studies through the application of multivariate statistical and computer-based taxonomic procedures. Finally some actual results taken from studies of archaeological materials of different types (ceramics, metals, glass *etc.*) will be reviewed as exemplifying the scope and value of this method.

2 Nuclear Activation of Archaeological Materials

Although the present essay is concerned with activation analysis by whatever nuclear reaction, and in fact will discuss research utilizing bremsstrahlung, fast proton, deuteron, and neutron bombardment, by far the lion's share belongs to 'thermal neutron activation analysis' or 'NAA'. Here the fraternity recognizes two subdivisions, depending on whether chemical separations after bombardment are or are not performed. In the latter case one speaks of 'INAA'—'instrumental neutron activation analysis', and most archaeological work has gone that route. The broad field of NAA has been extensively reviewed,[24] good textbooks exist,[25] and symposia of broad scope are periodically held.[26] Instead of ploughing that broad field once again, I would rather concentrate on those aspects of NAA specifically relevant to its archaeological applications.

As in all analysis, there are inherent questions of analytical precision, sampling, and natural distribution. Since the materials chosen by primitive man for his daily

[20] E. E. Richards, *Archaeometry*, 1959, **2**, 23.
[21] V. M. Emeleus, *Archaeometry*, 1958, **1**, 6; *ibid.*, 1960, **3**, 16.
[22] E. V. Sayre and R. W. Smith, *Science*, 1961, **133**, 1824.
[23] E. V. Sayre, 'Comptes Rendus 7th International Congress on Glass, Brussels, 1965', Gordon and Breach, New York, 1966, Paper No. 220.
[24] W. S. Lyon, E. Ricci, and H. H. Ross, *Analyt. Chem.*, 1974, **46**, 431R.
[25] D. De Soete, R. Gijbels, and J. Hoste, 'Neutron Activation Analysis', Wiley, New York, 1972, Chap. 10.
[26] 'International Conference on Modern Trends in Activation Analysis, Gaithersburg, Maryland, October, 1968', ed. J. R. De Voe and P. D. La Fleur, National Bureau of Standards, Special Publication 312 (2 volumes), 1969.

use and adornment came from nature rather than the factory, we would anticipate that even with excellent sampling and perfect laboratory analysis there would still be a 'natural' spread of analytical values within a particular, 'homogeneous', archaeologically meaningful class. This can be expressed as a 'natural' variance S_N^2. Inasmuch as we cannot possibly sample perfectly—*i.e.* we remove only a few pounds of obsidian from a flow of many tons, a few kg of clay from a metre-thick bed, and of these field samples take still smaller samples for laboratory analysis, there is also a sampling variance, S_S^2, to be added. And finally, our analytical procedures themselves are inherently imprecise, contributing an overall analytical variance we can call S_A^2. In activation analysis, the variance S_A^2 contains errors due to counting statistics, weighing, processing the gamma-spectrum, flux variations in the reactor or particle stream, decay and background corrections, contamination, and counting geometry. For our original, 'homogeneous' archaeological class we then observe a spread in observed analytical values

$$S_T^2 = S_N^2 + S_S^2 + S_A^2 \qquad (1)$$

This equation demonstrates that there is no point in working hard to achieve very precise analyses (S_A^2 small) in a system where the natural variance S_N^2 is large. The magnitude of S_A^2 in any laboratory analytical system is readily determined, element by element, by carrying out replicate analyses on a 'standard' substance, *i.e.* one which is similar to the archaeological material, but having no natural spread or sampling error. In principle one could estimate the sampling error and, consequently, the natural spread in an archaeological system. We can, however, survey the standard deviations associated with reported analytical values, $(S_T^2)^{1/2}$, and these will give us an idea of at least the maximum values of the natural widths, for various types of materials.

In the analysis at Brookhaven of groups of Greek ceramics of late Bronze age we found S_T values for many elements in the range 4—9%.[27] Since we knew the values of S_A for these same elements from standard analyses, we could conclude that for some elements, the observed variance S_T^2 was being contributed almost entirely by the analytical method. But for many others, the analytical method was 'good enough', *i.e.* not contributing significantly to the total spread. Perlman, Asaro, and co-workers at Berkeley have taken somewhat more pains to achieve higher analytical precision, and for groups of ceramics from the Middle East and Rome they report S_T values as small as 2—7%.[28,29]

These, however, may be exceptional cases. In Mesoamerican ceramics which we have analysed, total S_T values are commonly 10—20%. When tempering materials are added in the course of pottery manufacture, variances can become much larger: this is the case in many archaeologically interesting potteries.

In obsidian, where many analyses have been carried out by non-nuclear (*i.e.* X-ray fluorescence) techniques, it is interesting to note that groups having S_T values of 4 to 20% are common.[30—33] In other stones, such as turquoise and steatite (chlorite),

[27] A. M. Bieber, D. W. Brooks, G. Harbottle, and E. V. Sayre, *Archaeometry*, in press.
[28] F. Widemann, M. Picon, F. Asaro, H. V. Michel, and I. Perlman, *Archaeometry*, 1975, **17**, 45.
[29] I. Perlman and F. Asaro, *Archaeometry*, 1969, **11**, 21.
[30] D. E. Nelson, J. M. D'Auria, and R. B. Bennett, *Archaeometry*, 1975, **17**, 85.
[31] F. H. Stross, D. P. Stevenson, J. R. Weaver, and G. Wyld, in 'Science and Archaeology', ed. R. H. Brill, M.I.T. Press, Cambridge, Massachussetts, 1971, pp. 210—221.

the range is much larger,[34] although relatively few cases have been adequately studied. With objects of precious metal, coins, or other artifacts (such as bronze or copper implements), it is possible that one may often be dealing with refined, re-melted, alloyed, or debased metals that could not be expected to form 'natural' groups. Here one is more accustomed to hoping for analyses which reveal patterns of deliberate mixing to achieve a particular technical or economic goal. Nevertheless, in some cases patterns of trace elements may still yield valuable provenience or authenticity information, as in the case of Sasanian silver (see Chapter 4, below).

It will be demonstrated below that because of natural correlation of elements, the best taxonomic group is not necessarily the one having the smallest standard deviation. The existence of correlated elements also has a profound effect on the desirable level of analytical precision: the relationship suggested in equation (1) is misleading in cases of high correlation, as will become evident in the discussion of this effect.

In sampling, the quantity of material taken for analysis ideally should be sufficient that the variance introduced, S_s^2 [equation (1)], does not significantly increase S_T^2, the total variance. Thus the sample size is, not surprisingly, related to the homogeneity. In a majority of archaeologically interesting cases, however, the optimum cannot be reached: consider the problem of sampling jade, turquoise, or obsidian artifacts, bronze museum-pieces, silver bowls, or highly valuable ceramic vessels or figurines, which cannot be mutilated. In the end, with materials such as these, one samples as unobtrusively as possible, and analyses what one can get.

With archaeological pot-sherds the restrictions are less serious: after all, at many digs hundreds of thousands of sherds are turned up. At Brookhaven it has been found that, in sampling fine-paste pottery, it is adequate to drill at several points in the body of the sherd, avoiding surface or altered ceramic layers, to produce a combined sample of 200—400 mg. From this, after mixing, drying *etc.*, 40 mg is taken for analysis. In the case of highly tempered or very inhomogeneous pottery, such as the Mesoamerican 'Thin Orange', it was found empirically that initial samples as large as one gram were necessary: when these larger samples were ground and mixed, a 40 mg sample withdrawn for bombardment was adequately representative.[35] Perlman and Asaro have used samples of 100 mg.[29]

A special case arises when one is sampling a source of raw material used in antiquity: a clay bed, a copper mine, or obsidian flow. Obviously, as many samples should be taken from as many points as practicable, not only to give the best representative sample of the source, but also to permit the exploration through analysis of the subtle interrelationships of the elements present: the simple and multiple correlations. These constitute valuable data that cannot be obtained from a single sample. Of course, we wish to have an idea of the value of S_N^2 for each source.

To complete our discussion of sampling, it is worth spending a moment on the problem of contamination—*i.e.* the inadvertent admixture of non-sample material with the sample to be irradiated, or of any radioactive contaminants included with the sample during counting. At Brookhaven, we originally sliced ceramic wafers

---

[32] A. A. Gordus, J. B. Griffin, and G. A. Wright, in ref. 31, pp. 222—234.
[33] D. P. Stevenson, F. H. Stross, and R. F. Heizer, *Archaeometry*, 1971, **13**, 17.
[34] P. Weigand, P. Kohl, G. Harbottle, and E. V. Sayre, unpublished research; also see ref. 149.
[35] R. Abascal-M., G. Harbottle, and E. V. Sayre, unpublished research.

from sherds to be analysed by means of a diamond saw: this is excellent in avoiding contamination. Perlman and Asaro used a drill made of a sapphire rod,[29] which should certainly give a clean sample. Aspinall and Feather utilized a diamond-impregnated hollow core drill in sampling flint: the cores were then treated with aqua regia and rinsed to remove surface contamination.[36] We have recently resorted to this type of drill in sampling hard-rock copper ores such as malachite and chrysocolla. Turquoise and steatite are such soft stones that nearly any careful drilling technique will suffice. Precious metals such as Sasanian silver have been sampled by rubbing minute (50 µg) samples from the cleaned surface onto prepared quartz tubes: the silver is bombarded *in situ* and then dissolved from the quartz for processing.[37, 38] Microdrilling of precious metals is also employed and leads to a sample more representative of the whole object than the surface-rubbed streak.[38] Wyttenbach and Schubiger,[39] in sampling Roman lead, sliced samples with a steel knife, and then cleaned them with a diamond knife before rolling standardized strips 0.3 mm thick.

The present Brookhaven procedure for drilling samples from potsherds makes use of normal commercially-available tungsten carbide drills: these have the advantage of low cost and perform well, but occasionally small chips break off, which contaminate the sample.[21] Such contamination is, however, easy to detect through the observation of very large characteristic tungsten-187 gamma lines from the activated specimens.

Samples of archaeological material may be packaged for bombardment in any of several ways: wrapping in pure aluminium foil, sealing into pure quartz ampoules, or polyethylene heat-sealed tubing. With careful testing, a wrapping or encapsulating material can sometimes be found that is sufficiently pure, *i.e.* free of trace elements which can be activated, that the irradiated specimens can be counted without removal of the container: a very desirable simplification. We have found a source of pure silica tubing which suffices in this way for archaeological ceramics.[40]

In bombarding samples with neutrons, except in special cases where epithermal fluxes are desired, as in analysis for uranium, it is a general rule that the best location in the reactor is that having the highest flux of well-thermalized neutrons. The presence of a fast neutron component near the core of a reactor can lead to interfering reactions, *e.g.* $^{56}Fe(n,p)^{56}Mn$ can interfere with the measurement of small amounts of manganese *via* $^{55}Mn(n,\gamma)^{56}Mn$, whereas $^{54}Fe(n,\alpha)^{51}Cr$ could cause an error in the determination of chromium.[41]

Very high neutron fluxes, although attractive in enhancing the activity due to elements at extreme trace levels,[38] bring their own problems. The heat generated by neutron capture may increase temperatures to the point where containment fails: such failures have been occasionally observed in the bombardment of ceramic or copper-ore specimens in quartz envelopes at a flux of 5×10^{14}, though not at 2×10^{14}

---

[36] A. Aspinall and S. W. Feather, *Archaeometry*, 1972, **14**, 41.
[37] A. A. Gordus, in 'Methods of Chemical and Metallurgical Investigation of Ancient Coinage', Royal Numismatic Society, London, Special Publication, Vol. 8, 1972, p. 127.
[38] P. Meyers, L. Van Zelst, and E. V. Sayre, in 'Archaeological Chemistry', ed. C. W. Beck, (Advances in Chemistry Series 138), American Chemical Society, Washington, D.C., 1974, Chapter 3, pp. 22—34.
[39] A. Wyttenbach and P. A. Schubiger, *Archaeometry*, 1973, **15**, 199.
[40] Suprasil T-20, U.S. Fused Quartz Co., Fairfield, New Jersey.
[41] D. De Soete, R. Gijbels, and J. Hoste, in ref. 25, Appendix 3, Tables 1 and 2.

n cm^{-2} s^{-1}. Silver samples of 0.5 mg melted under a flux of 5×10^{14} n cm^{-2} s^{-1}, indicating a local temperature of nearly 1000 °C. Turquoise was in some cases totally destroyed (*i.e.* reduced to a black powder) in 2 h at 2×10^{14} n cm^{-2} s^{-1}. We have found, on the other hand, that heat-sealed polyethylene will withstand a well-thermalized flux of 1×10^{14} for several hours, in a water-cooled bombardment site, if the material being bombarded does not have high capture cross-sections. Clearly, careful trials are needed to establish routine bombardment conditions, whether with neutrons or charged-particle beams.

The bombardment of whole artifacts in order to achieve a non-destructive analysis poses several problems: one is the provision of a standard that has compositional and geometric properties sufficiently like those of the object to be analysed. Here there is, of course, no possible control over sampling. In such objects in a neutron flux one can also encounter the phenomenon of 'self-shielding', *i.e.* the surface layer's absorption of neutrons effectively lowers the flux available to the interior, leading to analytical values that are too low overall and that emphasize the surface composition. One can hope to overcome this effect by use of a proper standard, when possible. Self-shielding and other errors due to flux depression are discussed in a book by De Soete *et al.*[25]

One further effect which must be carefully checked if one is to bombard whole objects 'non-destructively' is colour change. In the intense radiation environments found in most reactors, objects of a crystalline or glassy nature will undergo radiation damage, resulting in the production of colour centres. For example, we found that turquoise exposed to a flux of 10^{14} n cm^{-2} s^{-1} for 10—20 min turned an unpleasant grey-green: the original blue could be only partially restored by thermal annealing at 110 °C after bombardment. Sigleo,[42] who employed much lower integrated fluxes, found no colour change, but of course needed proportionately longer counting times, for turquoise. One has every reason to expect that colour centres would also be formed in fast charged-particle bombardments. Metals do not show this effect, however.

Counting procedures involving Ge–Li detectors, feeding multichannel analysers whose output is dumped to magnetic tape or processed by an on-line computer, are now routine. Sample-changers coupled to these counters are very useful in that archaeological work almost necessarily implies large numbers of analyses, and, with counting times from one to a few hours per sample, permit one greatly to increase the output, by making use of night hours, holidays *etc.* for counting.

Computer methods for the analysis of gamma spectra are also routine. It is instructive, however, to check the computer output peak integrals against those obtained by hand calculation before deciding on a particular algorithm for spectrum analysis.

The crucially important question of standardization is double-pronged: first there is the choice of a standard, or the preparation of one in the laboratory, or the decision to operate with flux monitors; and then there is the technical point of making sure that the standard (or flux monitor) samples the same flux as the material whose analysis is being made.

There is a very practical reason why the standard chosen should chemically resemble fairly closely the sample analysed, in that it contain the same elements in

---

[42] A. C. Sigleo, *Science*, 1975, **189**, 459, and private communication.

roughly the same concentration range; at least, that the gamma-ray output of the standard differ not greatly from that of the sample. For every gamma counter, Ge–Li or NaI–Tl, there exists a maximum count rate at which the spectrum begins to deteriorate. If the irradiated standard emits too high a radiation level, then it and the samples it calibrates must be counted far away from the detector, with the result that the counting statistics of the samples are poor, and conversely if the samples are much more active than the standard. Of course, it is always possible to count the standards and samples at different distances, but this demands the introduction of additional (geometric) correction factors, and is impractical if sample changers are used.

There are nominally two kinds of standards: primary and secondary. If one prepares solutions containing appropriate and accurately determined concentrations of the elements desired, and accurately pipettes these solutions into a suitable container for irradiation, drying them therein while observing all necessary precautions, one probably can approximate the primary standard—*i.e.* the amounts are known, and the sampling and natural-spread variance terms in equation (1) are reduced to small values. However, it has been more common practice to take a large quantity of a carefully prepared, presumably homogeneous material such as clay,[43] pottery,[29] or glass[44] and first calibrate this by comparison with primary standards. This 'secondary' or 'in-house' standard is then used routinely for standardizing bombardments.

The U.S. Geological Survey many years ago prepared two standard rocks, a diabase (W-1) and granite (G-1), and these were analysed by many laboratories; their results have been statistically analysed to provide 'best values'.[45] More recently, the U.S.G.S. has provided six more standard rocks, which have also been repeatedly analysed by many different techniques.[46,47] There seems little question that for the major constituents, Si, Al, Fe, Ca, Mg, Na, K and Ti, the statistically treated mass of analytical data can be made to yield values which are essentially 'primary standard' in quality. However, with trace elements the picture is less favourable, and examination of Flanagan's tabulation of results coming from different laboratories is a chastening experience for one who may be enthusiastic for exact analysis of trace elements. Chase[48] has published a similar inter-laboratory comparison of analyses of several ancient bronzes: again, the results are not reassuring. The I.A.E.A. has also been responsible for several inter-laboratory comparison programmes involving only activation analysis.[49] Standard deviations are typically ± 15 to $\pm 30\%$ for a number of elements present at the p.p.m. level.

Flux monitors are simply reproducible, weighed bits of a substance easily handled and counted, for example metallic wire, whose measured activity is related to the integrated flux of neutrons at the bombardment site. In principle, if the cross-

---

[43] At Brookhaven, we are at present experimenting with the use of a commercial clay 'Ohio Red' supplied by Stewart Clay Co. (New York) as an 'in-house' standard. It has very desirable levels of trace elements: see the Table.
[44] E. V. Sayre and L.-H. Chan, in ref. 31, pp. 165—181.
[45] M. Fleischer, *Geochim. Cosmochim. Acta*, 1969, **33**, 65.
[46] F. J. Flanagan, *Geochim. Cosmochim. Acta*, 1969, **33**, 81.
[47] F. J. Flanagan, *Geochim. Cosmochim. Acta*, 1973, **37**, 1189.
[48] W. T. Chase, in ref. 38, pp. 148—185.
[49] J. Heinonen and O. Suschny, in 'Nuclear Activation Techniques in the Life Sciences', International Atomic Energy Agency, Vienna, 1971, Paper IAEA-SM-157/30, pp. 155—173.

sections, isotopic abundances, half-lives, detector efficiency factors *etc.* of the monitor isotopes, and of those counted in the sample, were known, it should be possible to calculate the analytical amounts of the different elements that were present in the unknowns, measuring only their gamma-ray peaks and those of the monitor. In practice, the monitors are used first with primary standards, with which they are irradiated and counted, to establish the calibrations, and afterwards on their own. Flux monitors are certainly convenient for routine measurements: they may give erroneous results if reactor neutron spectra should alter.

For best results, either standards or flux monitors should be placed immediately next to, or bracketing, the unknown samples. This need arises from the gradient in the flux, which can easily be a few percent per centimetre, at the bombardment site, and in any case ought to be determined. In bombarding with charged particles, samples are often 'sandwiched' between standards or monitors.[50]

The various laboratories which have been engaged in nuclear activation analytical studies in archaeology have adopted somewhat differing standardization procedures. At Brookhaven, our practice for several years has been to include samples of all six U.S.G.S. standard rocks with each bombardment: the values of each element in each rock are taken from a table,[51] which was in turn calculated from the raw data tabulated by Flanagan[46] through the application of Chauvenet's Criterion. After counting the standards, mean values of the calibration coefficient (counts per minute per milligram per unit concentration) are computed, and these mean values are then applied to the unknown samples. Other laboratories have employed 'in-house' standards such as pottery[29] or other materials calibrated against primary standards or primary standards themselves.

The problem of standardization is a difficult one, particularly at the trace level. Perhaps the best solution lies in the continual refinement of analytical techniques, constant checking against primary standards, and repeated inter-laboratory comparisons of reference material. It must be remembered that the ultimate goal is full interchangeability of comparable data, so that every laboratory engaged in archaeological research can with confidence make use of the published analyses of every other. We are far from that goal today, but getting appreciably closer.

3 Interpretation of the Analytical Data in Provenience Studies

Let us assume that the analyses of the pot-sherds and lumps of clay have been satisfactorily completed, and that we and the archaeologist with whom we collaborate are comfortably seated at a large table, examining the data. In this cosy scene I emphasize that both are present, for in fact the input which the archaeologist makes on questions of provenience, probable trade routes, stylistic matters, addition of temper, mineral inclusions *etc.* is crucial to a successful interpretation of the analytical data: the goal, after all, is to organize the sherds and clays into archaeologically meaningful groups. The archaeologist's contribution, of course, began much earlier with the formulation of the problem, asking the questions which it was hoped the analytical data would answer. An example or two will suffice.

In 1962 Professor Ignacio Bernal, Director of the National Institute of Anthro-

---

[50] P. Meyers, *Archaeometry*, 1969, 11, 67.
[51] R. Abascal-M., G. Harbottle, and E. V. Sayre, in ref. 38, pp. 81–99.

pology and History of Mexico, requested the Reporter's help in solving a problem in Mexican Precolumbian ceramics. There is a famous and beautiful polychrome pottery called 'Polychrome of Cholula' dating to the Mixteca–Puebla period, *ca.* 1300—1400 A.D., found not only around Cholula and through the Valley of Mexico, but in virtually identical patterns far to the south in Oaxaca, at sites such as Yagul. The first question was, then, whether one centre of manufacture supplied both north and south. To answer this a selection of sherds from sites in these two regions were selected and analysed. The answer was clear, that the sherds from the northern sites all had a common origin, and were very different in their paste composition from those found in Oaxaca. Since then, at least one of the Oaxacan polychrome sherds has been found to be very similar to other, presumably local, Oaxacan non-polychrome specimens.[51] A second question was equally clearly posed, and answered. Noguera[52] had recognized three stylistic groups of polychromes, the laca, maté, and firmé: did these three stylistic traditions also represent three different production centres? Abascal and Harbottle were able to decide unequivocally that all three came from one source, again the same one supplying the 'northern' or Cholula group mentioned above. At the start, clear hypotheses must be stated.

At the end, we confront the analytical data and ask the question, 'On the basis of paste composition, which sherds are similar: which will form statistically valid groups?' Since there are many, perhaps 20 to 30, elementary analytical concentrations for each sherd, we have passed willy-nilly from chemistry into Numerical Taxonomy, and must inevitably employ the techniques of multivariate analysis, in one form or another, to discover the structure in such a mass of data. Before investigating techniques for doing this, let us define what we have in such data, in accord with good archaeological usage.[53] The archaeologist defines the 'attributes' of an artifact in mutually exclusive descriptive terms such as colour, material, profile shape, mouth opening, while the values of these descriptive terms (green, flint, everted rim, 25 cm) are 'attribute states'. In our data, the attributes measured are the elements, for example iron, while the attribute states are the amounts, for example 8.73%. Although the discussion will focus on ceramics, for that is where most work has been done, the methods are applicable to data groups arising from the analysis of almost any sort of archaeological material, as will be mentioned in the next Section.

The Data Matrix and its Transformations.—Let us begin, then, by looking at the Table of analyses of some actual archaeological ceramics and a few clays from a variety of locations. We are immediately struck by two things: the very large range of elementary concentrations measured by nuclear activation as we go across a row, and the rather narrow ranges of particular elements as we go down the columns, comparing one sample with another. The first is easily explained: the concentration of an element detectable by NAA depends on nuclear properties such as isotopic abundance, cross-section, and half-life, and all these vary over orders of magnitude. The second observation is linked to clay mineralogy and the processes of formation of clay from rocks. Even so, it is astonishing that clays from such geographically

---

[52] E. Noguera, 'La Ceramica Arqueologica de Cholula', Editorial, Guarania-Mexico, 1954.
[53] J. E. Doran and F. R. Hodson, 'Mathematics and Computers in Archaeology', Edinburgh University Press, Edinburgh, 1975, p. 99.

Table Analyses of archaeological ceramics and clays: oxides in p.p.m. except where noted as per cent

Source	Na$_2$O %	K$_2$O %	Rb$_2$O	Cs$_2$O	BaO	Sc$_2$O$_3$	La$_2$O$_3$	Eu$_2$O$_3$	HfO$_2$	ThO$_2$	Ta$_2$O$_5$	Cr$_2$O$_3$	MnO	Fe$_2$O$_3$ %	CoO	Sb$_2$O$_3$
Mycenae; Greek Late Bronze Age[a]	0.60	3.31	163	8.47	427	35.1	39.4	1.62	3.48	13.0	1.23	326	1260	7.60	36	—
Classic Mayan; Guatemalan Mexican Y Fine Orange[b]	0.94	2.47	124	4.48	715	31.0	51.6	1.99	7.14	14.3	1.51	662	1170	7.74	40	0.72
Histria, Roumania; Greek Period[c]	1.80	2.52	107	4.36	543	19.3	44.6	1.56	7.47	13.3	—	131	863	4.53	15.2	1.25
Cholula, Mexico; Post-classic Polychrome[d]	2.44	1.02	58	2.91	634	23.8	—	1.45	5.00	6.05	0.86	198	897	6.14	20.7	0.54
Sakkara, Egypt; Archaic pottery[e]	2.28	1.80	70	1.2	410	37	46	2.5	6.7	9.2	2.6	236	2050	10.20	46	—
Babylon, Mesopotamia; pottery[f]	2.06	1.78	53	2.75	417	32.7	25	1.56	3.47	7.32	0.94	601	1390	7.33	36.8	—
Palestine; Red field clays[f]	0.86	1.62	48	1.46	860	21.0	37.9	1.78	9.93	8.56	1.84	152	954	5.69	24.2	0.94
Kition; Cyprus clay[g]	1.17	1.85	60	3.76	192	32.1	20	1.04	2.25	5.88	0.90	295	966	6.85	27	—
Oteapan; Mexican white 'cantaro' clay[h]	0.119	1.96	102	7.82	—	36.6	87	4.10	6.49	15.2	1.81	115	125	2.21	12.3	1.48
Ohio red clay, Stewart Clay Co.[i]	0.176	3.78	167	9.85	719	29.3	55.7	1.90	6.98	16.6	1.68	123	421	7.18	23.4	1.55

[a] Ref. 27. [b] R. L. Rands, R. L. Bishop, G. Harbottle, and E. V. Sayre, unpublished. [c] Ref. 54. [d] Ref. 55. [e] Ref. 56. [f] Ref. 57. [g] Ref. 58. [h] Prof. M. D. Coe, unpublished research. [i] Ref. 43.

[54] P. Alexandrescu, *Dacia*, 1972, **16**, 113.
[55] R. Abascal-M. and G. Harbottle, unpublished research.
[56] S. K. Tobia and E. V. Sayre, in 'Recent Advances in Science and Technology of Materials', ed. A. Bishay, Plenum Press, New York, 1974, Vol. 3, pp. 99—128.
[57] D. Brooks, A. M. Bieber, jun., G. Harbottle, and E. V. Sayre, in ref. 38, pp. 48—80
[58] A. M. Bieber, jun., D. W. Brooks, G. Harbottle, and E. V. Sayre, 'Proceedings International Conference on Application of Nuclear Methods in the Field of Works of Art, Rome and Venice, May, 1973', in press.

widely separated localities can have such similar compositions. In the Table only Na$_2$O and MnO show even one order of magnitude range. Had our Table included more pottery sources and more elements, we might have added Cr$_2$O$_3$ and CaO to the sodium and manganese, but would have also added TiO$_2$, NiO, Lu$_2$O$_3$, UO$_2$, Sm$_2$O$_3$, and Yb$_2$O$_3$ to the group generally showing less than an order of magnitude range. We may note that in the case of scandium the range 15—40 p.p.m. will include 80—90% of all clays and pottery the world of archaeology can offer. Given these small concentration ranges, and remembering that the variance terms in equation (1) arising from natural spread *etc.* can easily add up to observed group widths in the range ±10 to ±20%, it becomes intuitively clear that effective discrimination between groups will not in general be possible if only a few elements are analysed, and that in order to achieve discrimination, since we do not know *a priori* which elements may be most effective, we had better plan to analyse for as many elements as possible. This is in accord with Sneath and Sokal's[59] first axiom of Numerical Taxonomy 'The greater the content of information in the taxa of a classification and the more characters on which it is based, the better a given classification will be.' For us 'characters' are the elements analysed, *i.e.* the 'attributes' as defined above.

The first observation mentioned in connection with the Table, that variations are large from element to element in a given sample, imposes another constraint upon data-handling procedures as the researchers move from analysis to taxonomy. We turn to the second axiom of Sneath and Sokal[59] for guidance: '*A priori*, every character is of equal weight in creating natural taxa'. If we wish somehow to give equal weight, in carrying out a classification, to variation in elements present at the parts-per-million and at percent levels, four orders of magnitude different, then we can do one of two things. We can either standardize the data, for example from each concentration subtract the mean value and divide by the standard deviation, which yields for each element a group mean standardized concentration value of zero and a variance of one, or we can take the log of the concentration, and then use the logarithms for all subsequent manipulations.

There are fundamental statistical questions that intrude. To carry out statistical (probability) tests of group membership, one must assume that for single elements the group populations are distributed normally or as Student's *t* for small samples. If we convert first into logarithms for use throughout, then the appropriate distribution in Nature should have been log normal, as opposed to simple normal distribution for standardized concentrations. Investigations in forensic and geochemical analysis have favoured log normality for trace elements, but normality for major constituents.[60] It would appear at this time to be an open question. At Brookhaven we are currently examining the skewness and kurtosis of elementary distributions in large well-defined groups of clays and archaeological sherds. The preliminary results do not appear to favour decisively either type of distribution.

Examination of distribution plots of other workers, *e.g.* of trace impurities in copper ores[61] and of major components in Gallo-Roman Terra Sigillata pottery,

---

[59] P. H. A. Sneath and R. R. Sokal, 'Numerical Taxonomy', W. H. Freeman, San Francisco, 1973, p. 5.
[60] W. C. Krumbein and F. A. Graybill, 'An Introduction to Statistical Models in Geology', McGraw-Hill, New York, 1965.
[61] R. Bowman, A. M. Friedman, J. Lerner, and J. M. Milsted, *Archaeometry*, 1975, **17**, 157.

tends to confirm our view. In the latter case, however, the authors[62] state that 'we have never found any distribution which could have induced us to use a logarithmic pattern rather than a linear one . . .'. But it is true that for very narrow distributions the linear and logarithmic approach each other, and a very large sample population would be needed to decide between them. It is also true that the elements measured by Picon *et al.* tended to be 'major' rather than 'trace' in level. Going further, it also remains to be seen how seriously the classification schemes would be affected if one did choose the 'wrong' distribution. It may prove not to be of great importance to know the exact character of the 'natural' distribution. The advantages of various kinds of transformations are discussed by Sokal and Rohlf.[63]

There are other considerations governing the choice of an initial data transformation which must be made if we are to accord roughly equal weight to a given percentage change in all elements. Standardization procedures necessarily involve all the group members (in establishing the mean and variance) in determining the final numbers assigned for each element in each sample. Therefore standardization is appealing when one has a closed universe of samples. Employing an initial log transformation, on the other hand, files the data points away into an absolute co-ordinate system, as will be seen below, which in no way alters as new samples are added. This is somehow very appealing in archaeological work, where the whole emphasis has been on trade patterns, and the universe of clay and sherd samples, flints, obsidian blades *etc.* is ever-expanding as laboratories continue to analyse and record new material.

The Hypergeometric Distance Matrix.—We may imagine, then, a table of raw data conventionally arranged with element concentrations by columns and samples by rows, transformed to a similar table with log concentrations. If the reader will now imagine the transformed Table to be expanded to include several hundred samples, he will perhaps agree that the task of recognizing groups of chemically similar (obviously never identical) samples by 'eyeball' plus mental comparison would be formidable, exhausting, and prone to error. Fortunately, that is not necessary: there are well-established techniques of multivariate analysis and numerical taxonomy that will do the job on the computer. The starting point is the construction of a hyperspace (a conceptual space having more than three dimensions), with the number of dimensions being the number of elements determined, and with mutually orthogonal co-ordinate axes, one for each element, scaled off in log concentration units. The complete analytical data of each sample then correspond to a point in this hyperspace, and a group of samples having similar overall compositions will form a cluster of points. Clearly, the measure of 'distance' between points is important. Many types of distances have been calculated[64] but we will mention only a few which have found application in archaeological analytical work.

The most intuitively recognizable distance measure is Euclidean distance. Between two points A and B in ordinary x–y space the distance is given by the Pythagorean rule

---

[62] M. Picon, C. Carré, M. L. Cordoliani, M. Vichy, J. A. Hernandez, and J. L. Mignard *Archaeometry*, 1975, **17**, 191.
[63] R. R. Sokal and F. J. Rohlf, 'Biometry', W. H. Freeman, San Francisco, 1969.
[64] P. H. A. Sneath and R. R. Sokal, in ref. 59, pp. 121–128.

$$\mathrm{ED}_{A,B} = [(A_x - B_x)^2 + (A_y - B_y)^2]^{1/2} \qquad (2)$$

where A_x is the x-co-ordinate value of point A etc. This distance calculation is as readily made in a hyperspace of n dimensions, and is simply

$$\mathrm{ED}_{A,B} = \left[\sum_{i=1}^{n}(A_i - B_i)^2\right]^{1/2} \qquad (3)$$

Where A_i is the ith co-ordinate value of sample A, that is, log concentration of the ith element. Note that there are n squared terms under the radical, one for each element. In the event that the analytical data for an element i are missing in a particular sample J, then the usual procedure is to drop out all terms such as $(J_i - K_i)^2$ from the summation. However, the distances so calculated will then be too small, and certainly not comparable to distances calculated between samples having complete data. This problem is easily overcome by using the 'mean Euclidean distance'

$$\mathrm{MED}_{A,B} = \left[\frac{1}{m}\sum_{i=1}^{n}(A_i - B_i)^2\right]^{1/2} \qquad (4)$$

where m is the actual number of squared terms ($0 < m \leq n$) which have been summed. Note that, when logarithmic concentration co-ordinates are employed, the difference $(A_i - B_i)$ is the log of the ratio of concentrations of the ith element in samples A and B. By convention, natural logs are used. Since $\log_e(X) \approx X - 1$, equation (4) calculates, in effect, a root-mean-square average of the deviations of the ratios A_i/B_i from unity, giving the MED a readily understandable interpretation. Conforming to what has already been said about the range of standard deviation observed in natural archaeological groups, this would suggest that, within such groups, pairwise mean Euclidean distances ought to be in the range 0.050—0.200, especially for non-tempered wares.

Euclidean distance can be thought of as 'ruler' distance, or 'as the crow flies'. Another distance measure is 'mean character difference', 'Manhattan', or 'City-Block' distance,

$$\mathrm{MCD}_{A,B} = \frac{1}{m}\sum_{i=1}^{n}|A_i - B_i| \qquad (5)$$

so-called because one proceeds from point A to B not in a direct line but by 'walking' along the co-ordinates and summing the steps. At one time it was customary to compare or match the overall analytical compositions of two samples by superimposing their 'profiles'—i.e. plots of log concentration of the elements arranged in a particular order.[44, 51] Obtaining the best profile match is equivalent to minimizing the mean character difference. The technique of profile matching led to another useful mathematical procedure, the 'best relative fit'. Here the complete set of element (log) concentrations for a sample is shifted by a constant amount up or down (i.e., each concentration divided by the same factor) to bring profiles into the best possible coincidence. This procedure has a rational basis in that ceramic clays are often mixed with 'temper', such as broken shells or quartz sand, before the pottery is fabricated, and the concentrations of the elements measured are often much lower in the temper than in the clay. Therefore the temper tends to act as a simple diluent, and this 'dilution effect' can be compensated by the mathematical trick of 'best relative fit'.[19] To perform this fit one defines a factor F

$$F = \left[\prod_{i=1}^{n} (A_i/B_i) \right]^{1/n} \quad (6)$$

then divides each concentration (*not* log) value A_i by the factor to produce the adjusted set $A_i' = A_i/F$, which has the 'best relative fit' to the set B_i. The 'cos θ' and correlation coefficient r_{AB} measures of similarity (see below) can also be used under certain circumstances to compensate for dilution effects.

It is worth noting that the squared mean Euclidean distance, SMED, has also found application,[65] in the following way. In a cluster of points in hyperspace the average SMED calculated between all pairs of points is a good index of the within-cluster variance. If the goal of a clustering procedure is to produce groups in which this variance is minimized, then SMED is the appropriate measure with which to work. We will return to this in the section on clustering.

So far, the 'distances' calculated are examples of taxonomic 'dissimilarities'—the similarity decreases and dissimilarity increases as samples become more unlike. An example of a taxonomic similarity measure is the cosine of the angle formed by the vectors from the origin to points A and B in hyperspace.[66] It is calculated by

$$\cos \theta_{A,B} = \frac{\Sigma(A_i)(B_i)}{[(\Sigma A_i^2)(\Sigma B_i^2)]^{1/2}} \quad (7)$$

where the sums run over all n elements analysed. If the two points have the same vector then $\theta = 0$ and $\cos \theta = 1$. Cos θ can run from the 'identity' value of 1 to -1 for diametrically opposed points, but note that the two points need not be coincident in space, but merely lie in the same direction from the origin, to give $\cos \theta = 1$. This property can, under certain circumstances, allow one to compensate for dilution effects with cos θ.

It is interesting that, for standardized variables A_i and B_i, $\cos \theta_{AB}$ is identical with the simple correlation coefficient r_{AB}—the 'correlation' between two samples. Imagine a plot in which the x-axis applies to all the A_i values (log concentrations) and the y-axis all B_i. The correlation coefficient for the n plotted points (A_i, B_i) is then r_{AB}, and this, too, is a similarity measure also ranging from -1 to 1. Since the correlation coefficient as a similarity measure 'averages across variables' it has been criticized as having a doubtful philosophical basis. However, this is curious, since $\cos \theta_{AB}$ is readily visualized in an n-dimensional space–vectorial system.

We may then proceed to calculate, for example, in log concentration hyperspace, the distance from every point A to every other point B, C, . . ., that distance to be a measure of the lack of similarity between samples, with $D = 0$ for identical samples [equation (4)]. If there are P samples there will be $P(P-1)/2$ distances. It is convenient to think of them as arranged in a $P \times P$ matrix: of course, only one half of the matrix need be filled in, as the distance from A to B is the same as the distance from B to A. For a typical archaeological project P can easily be 500; the time to calculate the

---

[65] D. C. Olivier, 'AGCLUS, an Aggregative, Hierarchical Clustering Program, Dept. of Psychology and Social Relations', Harvard University, Cambridge, Massachussetts.

[66] F. J. Rohlf and R. R. Sokal, *Univ. Kansas Sci. Bull.*, 1965, **45**, 3.

distance matrix (for 20 elements) hand-punching it into a desk calculator would be well over a year, working 24 hours a day. The Brookhaven CDC-7600 will perform this task in 50 s, while writing all the 249 500 distances in the matrix on a disk for future use. Our computer program NADIST,[67] will also, on request, for each sample, sort the matrix distances to all other samples, so that by inspection one can find which samples are close to which others and hence form a cluster. This visual distance-clustering works well up to about a 20 × 20 matrix, but rapidly becomes tedious thereafter. It is, however, more quantitative and less subjective than visual profile-matching.

Cluster Analysis.—For large distance matrices one turns again to the computer, now programmed to perform cluster analysis; this is not a statistical technique *per se*, but rather a group of numerical procedures which aim to discover the structure inherent in the data. This means, of course, to map out the regions of hyperspace having significantly greater densities of points. One cannot do better than consult Chapter 5 of Sneath and Sokal,[59] which describes a whole range of clustering techniques.

We first consider a widely used approach, SAHN, which stands for sequential, agglomerative, hierarchic, non-overlapping clustering. 'Sequential' implies a recurring mathematical operation which gradually attacks the universe of sample points in hyperspace, either by dividing up ('divisive') or by linking together ('agglomerative') the points into clusters. In the latter case, one usually starts by considering each point to be a primitive cluster of one member, and links the closest points together, under a particular mathematical criterion (the 'clustering criterion') to proceed. 'Hierarchic' means that, as clustering proceeds, smaller clusters are merged to form larger ones with, however, decreasing affinity or cohesiveness, until the whole universe of points is one giant cluster, albeit often a loose one. 'Non-overlapping' means that, at a given level of clustering—*i.e.* at a point where the groups already formed have a specific tightness, or degree of relationship as defined by the clustering criterion, a point may not be contained in more than one group. Thus SAHN procedures have the property that, as clustering proceeds, points are swept into clusters and may not be moved around, until all points are used up.

Let us now follow the course of a SAHN cluster analysis. At first, if there are P points there are P clusters, each with one point. The $P \times P$ distance matrix is then searched, and the shortest distance found, to define the first 2-point cluster. The matrix is then searched again, and either two new points are joined, or a third point is added to the first two-point cluster, depending upon the value of a 'clustering criterion', which is a particular mathematical function of the distance that is evaluated in order to decide what happens next. There are many possible SAHN criteria: the program AGCLUS,[65] which we use at Brookhaven, contains seven different ones. Let us look at a few of them.

Single Linkage. This criterion, also called 'nearest neighbour', examines all remaining points and chooses, as a successful candidate to join a cluster, that point having the shortest distance to *any* member of the cluster. When two clusters A and B are being

---

[67] A. M. Bieber, jun., 'NADIST, a Program for Calculating Different Kinds of Taxonomic Distances', Brookhaven National Laboratory, Upton, New York.

joined to form a new cluster C, the shortest distance from *any* member a of A to *any* member b of B is decisive. Remember that single points are initially treated as clusters. The criterion is then, in shorthand,

$$d(\text{single linkage}) = \min d(a,b) \qquad (8)$$

Lance and Williams[68] have criticized single linkage as a strategy 'theoretically lacking in power'.

Complete Linkage. This criterion is also known as 'farthest neighbour' or 'longest link'. A candidate, to be admitted to a cluster, has the *shortest* distance to the *farthest* member of the extant cluster. For two clusters to be joined:

$$d(\text{complete linkage}) = \max d(a,b) \qquad (9)$$

It is interesting that whereas single-linkage clustering tends to produce long, skinny clusters ('chaining'), complete linkage leads to tight, discrete, and generally hyperspherical clusters.[69] The concept of a maximum distance leads us to one definition of 'size' of a cluster.

$$\text{size} = \max (a,a') \qquad (10)$$

i.e. the greatest distance, or 'diameter', between points a,a' in the same cluster.

Average Linkage and Centroid. In average-linkage clustering the arithmetic average of the distances is taken between the point and each member of the cluster, or for two clusters, between every possible pair of points made up of one in each cluster.

$$d(\text{average linkage}) = \frac{1}{n_1 n_2} \Sigma\, d(a,b). \qquad (11)$$

This cluster method is abbreviated UPGMA for unweighted pair-group method using arithmetic averages.

A variant of this is 'centroid clustering', in which the distance is taken from the candidate to the centroid of the cluster, or, in the case of two clusters, from centroid of A to centroid of B. The centroid of a cluster is its 'centre of gravity', easily found by taking the mean of the co-ordinates of the points in the cluster: thus

$$d(\text{centroid}) = d(\bar{a},\bar{b}) \qquad (12)$$

where $\bar{a}$ and $\bar{b}$ are the centroids and $d(\bar{a},\bar{b})$ the centroid-to-centroid distance, which is to be minimized.

Size-of-cluster Methods. A rather crude measure of the 'size' (*i.e.* hyperdiameter) of an irregularly shaped cluster was given in equation (10) above. A better way to define size is as the mean of the distances between all distinct pairs of items a and a' *within* the cluster A:

$$\text{size} = [2/n_A(n_A-1)] \sum_{a \neq a'} d(a,a') \qquad (13)$$

---

[68] G. N. Lance and W. T. Williams, *Computer J.*, 1966, **9**, 60.
[69] P. H. A. Sneath and R. R. Sokal, in ref. 59, p. 222.

where n_A is the total number of points in cluster A. It is possible to carry out the size calculation of equation (13), but to take into account *all* a and a', that is, to add in distances of points to themselves (by definition, zero) for dissimilarity measures and of point b from point a as well as a from b. Then

$$\text{size} = (1/N_A^2) \sum_{\text{all } a,a'} d(a,a') \qquad (14)$$

Although the procedure sounds rather odd, the size is a useful one in that it equals the mean-squared distance, or within-cluster variance of the points from the centroid, if the input distances are squared-mean Euclidean. This would appear to be a statistically sound and 'natural' measure of size. In all of these cluster methods based on size, the successful candidate for joining a cluster, or for merging one cluster with another, is that which produces the smallest new 'size'.

If the size based on equation (14) is weighted by the number of points in the cluster, so that

$$\text{size} = (1/N_a) \sum_{\text{all } a,a'} d(a,a') \qquad (15)$$

then the size becomes the within-cluster sum-of-squares.

One final system which we find produces excellent groups with analytical data from archaeological material minimizes the 'increase in size' criterion, defined as the size of the new cluster C minus the sizes of the merging clusters A and B. These sizes are defined by equation (15), using squared-mean Euclidean distance. The size-of-cluster methods described here are incorporated in the program AGCLUS.[65]

It is an interesting discovery of Lance and Williams[70] that the clustering criterion for strategies such as nearest-neighbour, farthest-neighbour, median, centroid *etc.* clustering can be obtained from a single, generalized, four-parameter equation. Although these parameters have definite values for the well-known clustering methods, Lance and Williams showed that, by arbitrarily varying them, an infinite range of clustering strategies, having any desired properties, could be generated.

Iterative Clustering. In the SAHN procedures described above, points, once assigned to clusters, must remain there. Clearly, it would often be desirable to move points around to produce improved, perhaps optimal, clusters. Such techniques are described as 'iterative clustering', and algorithms exist for carrying them out in the computer:[71] the programs are generally costlier in computer time than SAHN procedures.

An interesting and widely used technique is 'k-means' clustering: here, a preliminary set of k cluster centres are chosen (perhaps by SAHN clustering) and corresponding centroids are established. Then, distances from samples to centroids are calculated, incidentally requiring far less computer storage space than the original distance matrix. Finally, points are reallocated to give the best clusters, based on a minimization of the sum of squares of distances of points from their respective centroids. The recent book by Doran and Hodson contains a good discussion of the 'k-means' technique.[72]

[70] G. N. Lance and W. T. Williams, *Nature*, 1966, **212**, 218.
[71] H. P. Friedman and J. Rubin, *J. Amer. Stat. Assoc.*, 1967, **62**, 1159.
[72] J. E. Doran and F. R. Hodson, in ref. 53, pp. 180—184.

The Presentation of Cluster Analysis Results: Dendrograms.—The results of a cluster analysis are commonly presented as a line-printer output diagram called a 'dendrogram', which makes visible the hierarchy unearthed by the cluster analysis program. The dendrogram, or 'phenogram' (following Sneath and Sokal,[59] pp. 259 *et seq.*) is a tree structure in which the points and clusters are tied together at junctions whose distance, measured from left to right in the examples shown (see Figures 4 and 5 (pp. 57, 58), is the value of the dissimilarity, or clustering criterion. In this way the 'clustering' of the samples visibly emerges: groups and subgroups can be identified through their varying nearness of relationship. It ought to be mentioned that the dendrogram is necessarily an imperfect two-dimensional representation of the relationships existing in multivariate space: the 'cophenetic correlation coefficient' of Sokal and Rohlf 'is a measure of the agreement between the similarity values implied by the phenogram and those of the original similarity matrix'.[73] For archaeological ceramics this coefficient has been found to have values of 0.6—0.8, whereas 1.00 would constitute perfect agreement.

Correlation Effects, Group Properties, and Mahalanobis Distance.—The above discussion of clustering based on a Euclidean distance matrix contains an implication that 'natural' clusters (*i.e.* clays from a given geological setting, sherds from a homogeneous archaeological group, or obsidian from a particular outcrop) are hyperspherical in multidimensional space, or at least, that distances of nearest approach of cluster centres are large compared to within-cluster diameters or dimensions. For example, in 'centroid' or 'k-means' clustering methods, the criterion is the simple distance from the centroid, in whatever direction, the candidate having the shortest distance being the next to join the group. But it is well known that in Nature two or more elements are often correlated[74, 75] and we have observed the same effect in archaeological samples.[76] A very striking case has been reported by Bowman *et al.*[77] in obsidian from Borax Lake: here the elements iron, scandium, thorium, caesium, manganese, and cobalt were multiply correlated with correlation coefficients of essentially unity.

In Figure 1 data from Brookhaven work, involving clays and pottery from the Middle East, are presented.[56, 57] The co-ordinates are concentrations of iron and scandium on log scales. The three groups are A red field clays, B limestone (hill) clays from a variety of sites in Palestine, and C 'Nile Mud' clays, and pottery made therefrom. The correlation coefficients are all close to unity, but note that the three groups do not all fall on the same correlation line. If we were to attempt to distinguish group A from group B on the basis of Fe and Sc, it is obvious that, if only the separate univariate distributions were considered, the two groups would hopelessly overlap with the mean Fe or Sc values in A and B differing from each other by only a fraction of one standard deviation. And yet it is obvious by inspection of Figure 1 that A, B, and C are all good groups. In fact, the ellipses drawn round the groups enclose regions of 95% confidence of group membership.

It is often observed, in plots like Figure 1, that the correlation line has a slope of 1

[73] P. H. A. Sneath and R. R. Sokal, in ref. 59, p. 278.
[74] A. W. Moore and J. S. Russell, *Geoderma*, 1967, **1**, 139.
[75] K. K. Turekian, A. Katz, and L. Chen, *Limnology and Oceanography*, 1973, **18**, 240.
[76] G. Harbottle, *Archaeometry*, 1970, **12**, 23.
[77] H. R. Bowman, F. Asaro, and I. Perlman, *Archaeometry*, 1973, **15**, 123.

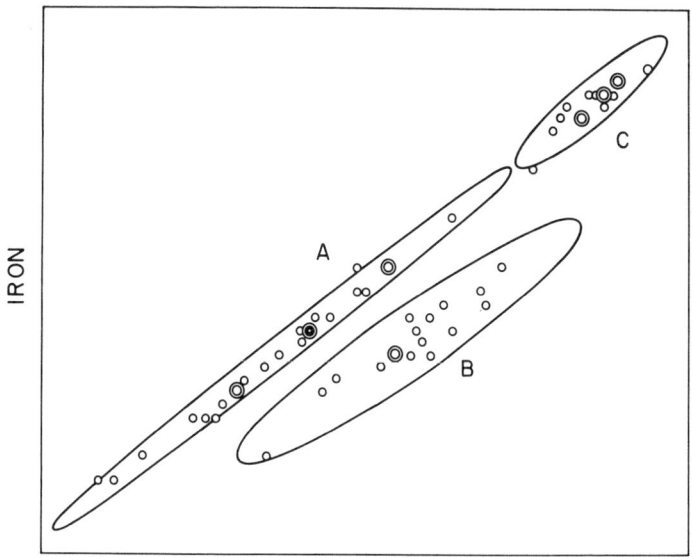

Figure 1 *Iron–scandium plot (arbitrary log scale, equal logarithmic increments). A = red field clays, B = limestone hill clays (Palestinian sites), C = 'Nile Mud' clays and pottery*

(45° angle) when the x and y co-ordinates give equal logarithmic increments—*i.e.* are scaled equally. The simplest geological interpretation is that elements x and y enter in a fixed proportion in one mineral, which is then diluted to varying degrees by other clay minerals which do not contain very much x or y. The knowledge of this proportion, which is easily obtained mathematically, might conceivably be of use in the numerical taxonomy of clay sources.

In the analysis of archaeological material we have found not only single but multiple correlation to be very common: here, a group of elements, often four or five, all show relatively high mutual correlation coefficients when taken two at a time. The obsidian data of Bowman *et al.*[77] already cited are another good example. Returning to the data in Figure 1, but imagining them to be extended to more dimensions, the ellipses which enclose the bivariate distributions at the 95% confidence level become ellipsoids—cigar-shaped surfaces—in three- and hyperellipsoids in many-dimensional space. It is clear that, if the situation portrayed in Figure 1 were present in n-dimensional space so that all the $n(n-1)/2$ two-dimensional plots were similarly stretched out along correlation lines, the SAHN clustering techniques based on Euclidean distance or mean character difference would in general not work well at 'finding' the clusters. However, this extreme case, although it is just what was found by Bowman *et al.* in one obsidian flow, does not seem to appear in ceramic clays. Here the situation, typically, is that for a 'good' group analysed for 20 elements only 10 or 20 out of 190 correlation coefficients will be larger than 0.8.

Thus, the great bulk of the data is not highly correlated, and since distance measures such as equations (2)—(5) average over all the data, we are probably safe in using Euclidean measures to find clusters. Solomon[78] says that 'As a rule of thumb, correlations as high as 0.5 will not produce Euclidean distances that lead to operational difficulties'. Parenthetically, the reader may wonder whether high negative correlation coefficients are observed: the answer is yes, in the obsidian data cited, and also, very occasionally, in archaeological ceramics, especially in pairs in which calcium is one of the elements.

In cases of serious multiple correlation, with several, perhaps partially overlapping, hyperellipsoidal clusters, the problem facing numerical taxonomists, namely, to discover this structure through the application of computer algorithms to the mass of initial data, or to the distance matrix, is indeed a formidable one. However, one can always begin with two-dimensional 'scatter diagrams' like Figure 1, with the points coded as to archaeological sites, or other geographic or cultural references. The computer of human eye plus brain is not at all bad at detecting and sorting out partially correlated data clusters, even with some overlap, if it is given some auxiliary clues.

If by eye, or through SAHN clustering, we can find groups, then there exists in principle a straightforward way of determining their statistical validities, that is, given a group, the rigorous calculation of probability of group membership of each member. This is equivalent to defining the hypersurface that encloses the points in n-dimensional space with a given confidence level, say 95%. What is involved is the calculation of Mahalanobis distance D^2,[79] which is simply the squared (standardized) Euclidean distance from the centroid to the point divided by the group standard deviation *in that direction*, *i.e.* along that line. The calculation is

$$D^2_{A,a} = \sum_{i=1}^{n} \sum_{j=1}^{n} (A_i - \bar{A}_i) W_{ij}^{-1} (A_j - \bar{A}_j) \tag{16}$$

where $D^2_{A,a}$ is the Mahalanobis distance from point A, having concentrations of elements i and j equalling A_i and A_j to the centroid a, representing mean concentrations $\bar{A}_i$ and $\bar{A}_j$, and W_{ij}^{-1} is the ith–j^{th} element of the inverse of the variance–covariance matrix for the group. The summation is carried out over all n elements, or dimensions.

In vector notation,[78] the (squared) Euclidean distance between individuals A and B is given by

$$\text{SED}_{A,B} = (A-B)'(A-B) \tag{17}$$

where A and B are column vectors each with n rows containing the data on individuals A and B, and the difference row vector $(A-B)'$ multiplies its transpose $(A-B)$, to give the scalar (squared) distance. Similarly, the Mahalanobis distance between the same individuals is

$$D^2_{A,B} = (A-B)' W^{-1}(A-B) \tag{18}$$

---

[78] H. Soloman, in 'Mathematics in the Archaeological and Historical Sciences', (Proceedings of the Anglo-Roumanian Conference, Mamaia, 1970), ed. F. R. Hodson, D. G. Kendall, and P. Tautu, Edinburgh University Press, Edinburgh, 1971, p. 67.

[79] P. C. Mahalanobis, *Proc. Nat. Inst. Sci. India*, 1936, **2**, 49.

where W^{-1} is the inverse of the variance–covariance matrix. Equations (16) and (18) point up the importance of the variance–covariance matrix in distance calculations: in this matrix resides all the information on the correlation of two or more elements in a group of analyses. Note especially that W^{-1} is the property of a group: it really attains maximum significance after a homogeneous group has been formed, and for this reason the Mahalanobis distance is especially useful for testing SAHN-(or otherwise) clustered groups.

If the group under consideration is multivariate normal (and of infinite size) then the Mahalanobis distances from the centroid will be distributed as chi-squared, with n degrees of freedom. If the group is a statistically random sample drawn from the infinite, multivariate, normal population, the probability of a given value of $D^2_{A,a}$ may be obtained from Hotelling's T^2, the multivariate equivalent of Student's t. Note that D^2 can be calculated between points [equation (18)] or between points and a centroid [equation (16)], hence can be used in clustering procedures in various ways, provided one settles the question of which variance–covariance matrix to use. Sometimes what is done is that all the data are 'lumped' to give a variance–covariance matrix that cuts across groups, but one probably obtains the best discrimination by basing the matrix on the data in one specific group.

At Brookhaven we use the program ADCORR[80] to calculate the correlation and variance–covariance matrix (and its inverse), Mahalanobis distance, probability of group membership for assumed multivariate normal groups, and characteristic co-ordinates (or vectors) giving new, uncorrelated axes which are linear combinations of the original log concentration axes, and to plot histograms of the sample distributions on these characteristic vectors. The characteristic vectors are similar to 'principal components', which will be discussed below, the difference being that, whereas principal components are derived from the correlation matrix, characteristic vectors derive from the variance–covariance matrix.[81]

To illustrate the advantage of the use of characteristic vectors and Mahalanobis distance in the discrimination of groups involving correlated elements, let us again turn to the data of Figure 1. The red field clays (group A) are used to define the variance–covariance matrix; the two axes log[Fe] and log[Sc] are then rotated orthogonally to form two new characteristic vector axes, a vector of greatest variance, along the direction of the semimajor axis of the ellipse of group A, and a vector of least variance, at right angles, along the semiminor axis of A. The origin of the new characteristic co-ordinate system is moved to the centroid of group A. Figures 2 and 3 show the distributions of specimen characteristic co-ordinates in the new system of axes, *i.e.* along the axes of greatest (Figure 2) and least (Figure 3) variance. The black, grey, and open histograms represent groups A (red field clays), and B (limestone hill clays), and C ('Nile Mud' clays and pottery), respectively, projected upon the two characteristic vector axes. Note that, although the characteristic vector of greatest variance effectively separates groups A and C, we must resort to the vector of least variance to separate A and B. We feel that this phenomenon—the ability of vectors of low variance in the transformed co-ordinate system to produce in some cases good discrimination between groups—

---

[80] E. V. Sayre, ADCORR, Brookhaven National Laboratory, Upton, New York, 1973.
[81] W. W. Cooley and P. R. Lohnes, 'Multivariate Data Analysis', Wiley, New York, 1971.

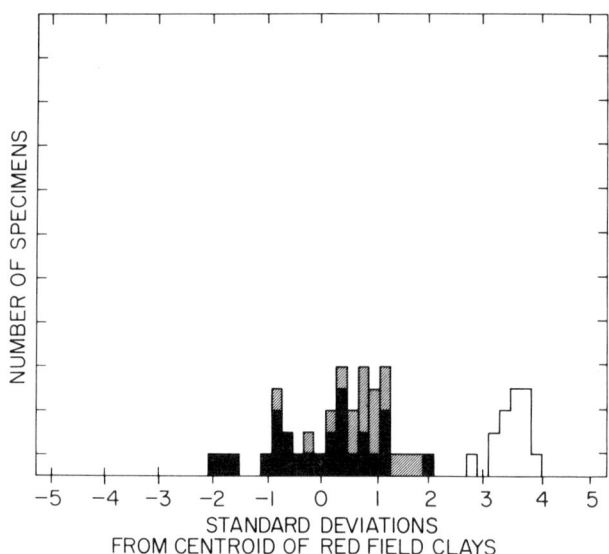

Figure 2 *Distribution of specimen co-ordinates for* Fe–Sc *characteristic vector of greatest variance. Black histogram = group A; grey = group B; open = group C*

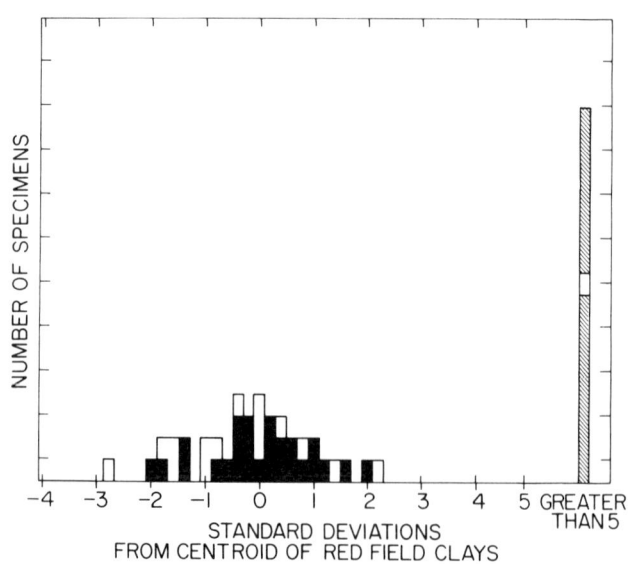

Figure 3 *Distribution of specimen co-ordinates for* Fe–Sc *characteristic vector of least variance. Black histogram = group A; grey = group B; open = group C*

has not been sufficiently emphasized in the literature, although Doran and Hodson,[82] in describing the work of Newton and Renfrew[83] on the Stone and Thomas faience bead analyses,[84] *do* note that 'the need for two sets of discriminant functions is interesting: some groups are split off basically by different discriminators than others'.

In the use of principal component analysis (see below), which is closely related to characteristic vectors, it is quite commonly the practice in showing clustering to plot only the first and second components out of many, or build 3-D models based on the first, second, and third components because these carry the great bulk of the variance. But the final components, with only a trivial fraction of the total variance, may nonetheless possess great discriminant ability, as noted above.

If the Mahalanobis distances, whose calculation is governed by the variance–covariance matrix of group *A*, be calculated, measured from the origin (centroid of group *A*) and then average-linkage SAHN clustering be carried out on these distances, the dendrogram in Figure 4 is obtained, showing the clean separation of the samples

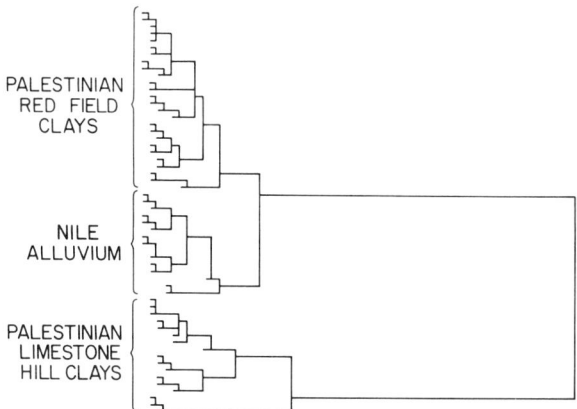

Figure 4 *Dendrogram of average-linkage clustering on Mahalanobis distances, based on group A (red field clays); data of Figure 1*

into the original three groups. This dendrogram may be compared with that of Figure 5, where the same type of clustering, but based on the original mean Euclidean distance matrix, was performed, and obviously was unable to separate groups *A* and *B*.

One severe problem which arises from the use of Mahalanobis distance (actually from the variance–covariance matrix) and the calculation of probability of group membership is what we have termed 'stretchability'. If the number of dimensions (elements measured) is n, and the number of samples P, the variance–covariance matrix will be singular and no inversion can be carried out, unless $P \geqslant n+1$. To illustrate the problem of stretchability, let $P \approx 2n$, and consider the volume enclosed

---

[82] J. E. Doran and F. R. Hodson, 'Mathematics and Computers in Archaeology', in ref. 53, pp. 251—257.
[83] R. G. Newton and C. Renfrew, *Antiquity*, 1970, **44**, 199.
[84] J. F. S. Stone and L. C. Thomas, *Proc. Prehist. Soc.*, 1956, **22**, 37.

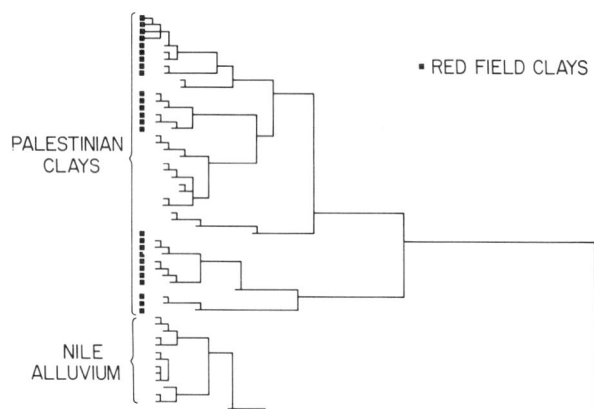

Figure 5 *Average-linkage clustering, data of Figure 1, based on mean Euclidean distances*

by the hyperellipsoid of 95% confidence. That volume is defined by the variance–covariance matrix of the P samples: if a $(P+1)$th sample lies outside that volume, and its probability of group membership be calculated using Hotelling's T^2, it would be found to be quite low, say 1%. But now arbitrarily place the 'outsider' *in* the group, and let a new variance–covariance matrix and a new 95%-confidence hyperellipsoid be calculated for the $P+1$ samples. The Mahalanobis distance from the centroid to the $(P+1)$th sample will now correspond to a quite *high* probability of group membership, say 80%: in other words, if P is not much larger than n, the hyperellipsoid is easily stretched to accommodate even a rather unlikely newcomer. How large need the ratio P/n be? Theoretical studies indicate that, as a rule-of-thumb, P/n should be at least 3—5, and the larger the better.[85] This P/n problem infects all analyses based on the variance–covariance matrix.

The use of Mahalanobis techniques has begun to appear in the literature of archaeological analysis: recent papers deal with obsidian[86] and ceramics.[27] The work of Newton and Renfrew on the faience beads originally analysed by Stone and Thomas by emission spectroscopy produced interesting groupings despite the rather roughly quantitative nature of the input data,[82] using discriminant functions.

Principal Components Analysis.—The correlation matrix R is the $n \times n$ matrix of correlation coefficients of the type described in reference 78: it describes the degree of interrelation between all the pairs of elements measured, and may be calculated[87] as

$$R = [1/(P-1)]XX' \qquad (19)$$

where X is the standardized data matrix for P samples and X' its transpose. If the characteristic equation of the correlation matrix R

---

[85] D. H. Foley, *I.E.E.E. Trans. Inf. Theory (USA)*, 1972, IT-18, 618.
[86] G. K. Ward, *Archaeometry*, 1974, **16**, 41; *Amer. Antiq.*, 1974 **39**, 473.
[87] P. H. A. Sneath and R. R. Sokal, in ref. 59, p. 245.

$$(R - \lambda I)v = 0 \tag{20}$$

be solved, the n roots $\lambda_1 \ldots \lambda_n$ are the eigenvalues and the corresponding vectors $v_1 \ldots v_n$ the eigenvectors. I is the identity matrix. The eigenvectors are orthogonal and form a new set of uncorrelated axes known as 'principal' axes. As in the case of the characteristic vectors mentioned above, the co-ordinates of these new axes are linear combinations of the original variables, but chosen in such a way that the first new 'principal' axis, having vectorial direction v_1, is the axis along which the greatest amount of variance[88] in the sample set lies—in other words, the axis that corresponds to the long dimension of the cigar-shaped ellipsoid. The first eigenvalue λ_1 is in fact equal to that variance. The second principal axis, orthogonal to the first, explains the second largest amount of variance, and so on. In a totally uncorrelated sample set, all these n eigenvalues (variances) would be the same and there would be no advantage to this procedure. But with correlated elements, a single new principal axis may largely account for the variance previously encountered in several elements, for example in a group of interrelated elements like scandium, lanthanum, cerium, and europium, or potassium, rubidium, and caesium. It often happens that the first two or three principal axes can account for 70—80% of the total variance in a sample set in a space of 20 dimensions: thus, the multidimensional space relationships can be telescoped or economically reduced into a space that the human mind can comprehend, namely, a space of two or three dimensions. This is, of course, of great help in the presentation of data. But, as was already noted in the section on Mahalanobis distance above, the vectors accounting for the greatest variance do not necessarily produce the greatest discrimination, simply because some of the elements which are good discriminators may be heavily 'loaded' on vectors of low variance.

The attack on the data and correlation matrices can be carried further *via* Factor Analysis, which has been used in one extensive study in archaeological NAA.[89] The interested reader is referred to a review by Klovan,[90] on Q- and R-mode factor analysis. The expression Q-mode refers in general to studies of the relationships between samples, while R-mode refers to the relationships between attributes, in this case element analytical concentrations, for a given set of samples. Thus correlation is an R-mode, and distance a Q-mode relationship.

Computer Programs.—Although some scientists prefer to write their own programs, making use of in-house subroutines, for example for matrix inversion, there are in fact many available programs for multivariate analysis, statistics, and taxonomy. Mention should be made of the very comprehensive program NT-SYS of Rohlf *et al.*[91] and its miniaturized version MINT. Then there are the SPSS[92] and BMD[93]

---

[88] Properly speaking the variance of the correlation matrix, which is based on standardized data. If the operations expressed in equation (20) are carried out as in the program ADCORR,[80] on the unstandardized data matrix, then the variance–covariance matrix is the analogue to the correlation matrix, the eigenvalues $\lambda_1 \ldots \lambda_n$ are true data variances, and the eigenvectors v the 'characteristic vectors' whose use was described in the preceding section.
[89] R. L. Bishop, Ph.D. Thesis, Southern Illinois University, Carbondale, Illinois, 1975.
[90] J. E. Klovan, in 'Concepts in Geostatistics', ed. R. B. McCammon, Springer-Verlag, New York, Heidelberg, Berlin, 1975, pp. 21—69.
[91] F. J. Rohlf, J. Kishpaugh, and D. Kirk, 'NT-SYS Numerical Taxonomy System of Multivariate Statistical Programs', State University of New York, Stony Brook, New York, 1971.

statistical packages. We have already spoken of NADIST,[67] AGCLUS,[65] and ADCORR.[80] The textbook of Cooley and Lohnes[81] contains numerous programs in FORTRAN for multivariate statistical analysis, including Mahalanobis distance. Taxonomic packages referred to in Sneath and Sokal[94] include those of Wishart, Gower, and Ross at Rothamstead, and Lance and Williams at Canberra. In the U.S.A., the group at the Kansas State Geological Survey has been particularly active and has documented many programs.

Summary.—A possible sequence of steps in the numerical taxonomy of archaeological samples, whose proveniences and analyses are known, but whose chemical relationships to one another are unknown, might be as follows:

(1) Conversion of analytical data to log data, unless range is about the same for all elements. Possible conversion to standardized data.

(2) Distance matrix calculation, using mean measures if there are any gaps in the analytical data.

(3) Clustering by several algorithms: comparison of clusters so obtained. Clustering by inspection for small ($P < 20$), sorted distance matrices. Calculation of simple statistical properties [mean, mean $\pm$ values corresponding to (*a*) standard deviation and (*b*) 95% confidence] of clusters so determined. Comparison of cluster overlaps element-by-element, in terms of standard deviations.

(4) Preparation of correlation matrices for any readily distinguishable large subgroups of the total data. Necessary to be sure correlation coefficient is significant for number of samples taken.

(5) Preparation of two-dimensional scatter diagrams, choosing both (*a*) pairs of elements that are highly correlated and/or (*b*) pairs of elements which show a considerable overall range of variation relative to standard deviation *within* any well-defined clusters. Plots can be labelled two ways: (*a*) by cluster, as found in (3) above, and (*b*) by archaeological site.

(6) Testing the validity of the final groups by Mahalanobis distance procedures, after reducing dimensionality to the point where $P/n \geqslant 3$, if that is necessary. Refinement of groups on the basis of Mahalanobis distance, or through iterative clustering, *k*-means *etc.* if groups are well-established.

To return to the scene with which this section began, it appears that the chemist and archaeologist have been joined at the table by a statistician, and the twofold interdisciplinary nature of the investigation has been expanded to three, to include multivariate statistics and numerical taxonomy.

4 Application of Nuclear Techniques to Specific Classes of Archaeological Material

In this section will be mentioned some of the analyses of archaeological material by nuclear techniques. The list is by no means exhaustive; a few cases where non-

[92] N. H. Nie, C. H. Hull, J. G. Jenkins, K. Steinbrenner, and D. H. Bent, 'SPSS Statistical Package for the Social Sciences', 2nd Edn., McGraw-Hill, New York, 1975.
[93] 'BMD Biomedical Computer Programs', ed. W. J. Dixon, University of California Press, Berkeley, 1970.
[94] P. H. A. Sneath and R. R. Sokal, in ref. 59, Appendix B, pp. 481—487.

nuclear methods were used are included if they seem of particular interest. Previous reviews of the field include those of Sayre,[95] Perlman, *et al.*,[96] Sayre and Meyers,[17] and Gordus.[97] Since the annotated bibliography of Sayre and Meyers covers the field very well from its inception through to 1971, the present chapter will emphasize reports published from 1972 onwards, including work in progress at the present time.

It is worth noting that many analytical studies of archaeological objects are published in the journal *Archaeometry*, the Bulletin of the Research Laboratory for Archaeology and the History of Art, at Oxford. The same laboratory arranges a three-day symposium every winter, which has become a unique meeting-ground for all those interested in the applications of modern scientific methods to archaeology. The *Journal of Radioanalytical Chemistry* also carried reports in the field, and the *Art and Archaeology Technical Abstracts*[17] provides excellent coverage of all published articles, books *etc.* The recently established *Journal of Archaeological Science* also carries analytical studies.

Ceramic Provenience Studies.—It was originally felt that if one analysed for enough elements, accurately enough, the assignment of particular sherds to specific clay sources might be made unequivocally. It is now generally realized that the best one can do is to give a probability for assignment or grouping because (*a*) as pointed out (Table) above, of the totality of n-dimensional hyperspace, all the ceramic clays in the world appear to occupy a relatively restricted, rather crowded volume, and (*b*) the addition of temper to clay, commonly practised, serves drastically to blur the definition of clusters in hyperspace. However, even a probability of identification can be valuable to the archaeologist, especially when he combines it with attribute data of shape, colour, decoration *etc.* Also, in many cases one may say with very high probability that a particular sherd *could not* have belonged to a certain group. Finally, it is often valuable to an archaeologist to be able to group sherds by source rather than by style, even though the source itself cannot possibly be found.

At the Fourth A.C.S. Symposium on Archaeological Chemistry in 1968, published in 1971,[31] there were several papers on ceramic provenience study. Perlman and Asaro[98] reported their analytical technique, with its emphasis on precision and the reduction of analytical error, their preparation of an 'in-house' pottery standard (which has since been distributed to a number of laboratories), tests of reproducibility, and finally the analysis of some archaeological sherds from Egypt ('Nile Mud') and from the Ashdod region of Israel. The authors raised the problem of the statistical treatment of multivariate data, assigning an 'index of disagreement' to samples on the basis of individual element distributions. As pointed out in Section 3, the existence of strong bi- (and multi-)variate correlation unfortunately affects the validity of judgements of non-group membership based on the assumption of independent univariate normality. Perlman and Asaro also pointed out the importance of 'dilution factors' for quartz-tempered pottery.

At the same symposium, Olin and Sayre discussed their results on English and

---

[95] E. V. Sayre, *Adv. Activation Analysis*, 1972, **2**, 156.
[96] I. Perlman, F. Asaro, and H. V. Michel, *Ann. Rev. Nuclear Sci.*, 1972, **22**, 383.
[97] A. Gordus, *Phil. Trans.*, 1970, **269A**, 165.
[98] I. Perlman and F. Asaro, in ref. 31, pp. 182—195.

American pottery of the Colonial period,[99] analysing the groups in terms of Student's t tests of univariate distributions, and 'best relative fit'. Of special interest were two sherds found at Drake's Bay, California: they were compared to similar material from North Devon, England, but had very different compositions.

Although the analyses were by emission spectrography, the results given by Hall[100] were a good example of the extremely far-ranging studies of Greek pottery by the Oxford group:[101] their clustering technique has been based on visual profile-matching but their data may be reworked with numerical taxonomy, as by Prag et al.[102]

Some of the sherds from Mycenae and Knossos at Oxford were reanalysed by Harbottle,[76] who found that the Oxford classification into profile-matched groups persisted when one moved to the level of trace elements. This same material, plus additional sherds from Greece, has recently been subjected to intensive chemical and statistical analysis by Bieber et al.,[27] and, with the application of more modern numerical taxonomic methods, the Oxford 'Composition A' material has now been split cleanly into separate Laconia and Berbati subgroups. Mycenaean ware, designated 'III C 1', found at Tell Ashdod has also been studied by the Berkeley group:[103] here is a good example of the possibility of cross-utilization of data between different groups, provided standardization is reproducible, or can be exchanged.

Pottery from other Greek colonies has also been investigated; Harbottle has determined that wares found at Histria, Roumania, could have been locally made.[54] At Brookhaven, Fillieres is currently engaged in the detailed investigation of amphorae coming from the Greek period at Marseilles.

In the Middle East region Davidson and McKerrell have obtained some extremely interesting data on trade patterns in Halaf ware, the sites being located in Northern Iraq and Eastern Syria. In the course of this work they analysed clays in a broad-ranging survey of local wadis and came to the surprising but significant conclusion that in some cases the chemical composition was the same today as in Chalcolithic times.[104] Working with sherds from Tell el-Hesi in Israel, and with many clays and sherds from other Palestinian archaeological sites, Brooks et al.[57] have been able, similarly, to correlate local field clays with pottery from early Bronze to Arabic periods; some results have already been mentioned above in the section on taxonomy (Figures 1 to 5). The Tell el-Hesi pottery includes import sherds from Greece and Cyprus: the study of locale of origin of Cypriote wares has been carried forward in two substantial projects, covering several archaeological periods, by Artzy, working in Berkeley,[105] and Bieber, at Brookhaven.[106] Returning

[99] J. S. Olin and E. V. Sayre, in ref. 31, pp. 196—209.
[100] E. T. Hall, in ref. 31, pp. 156—164.
[101] The Oxford Laboratory's analyses of large numbers of Greek ceramics have, for the most part, been published in *Archaeometry*, and in the *Annual of the British School of Archaeology at Athens*.
[102] A. J. N. W. Prag, F. Schweizer, J. Ll. W. Williams, and P. A. Schubiger, *Archaeometry* 1974, **16**, 153.
[103] F. Asaro, M. Dothan, and I. Perlman, *Archaeometry*, 1971, **13**, 169.
[104] T. E. Davidson and H. McKerrell, 1975, personal communication.
[105] L. M. Artzy, Ph.D. Thesis, Brandeis University, Waltham, Massachussetts, 1972 (University Microfilms Dissertation Abstr. 72-32085); L. M. Artzy, F. Asaro, and I. Perlman, *J. Near Eastern Studies*, in press.
[106] A. M. Bieber, jun., Ph.D. Thesis to be submitted to Dept. of Anthropology, University of Connecticut, Storrs, Connecticut; also see ref. 58.

to the mainland, the work of Asaro and Perlman on Mycenaean wares is mentioned above; additional data on other wares from Tell Ashdod have also been published,[98] and Tobia and Sayre have made extensive comparisons of Egyptian clays and archaeological pottery.[56] Clay sources of North Africa which may be related to Islamic pottery are currently being surveyed by a group affiliated with the Metropolitan Museum of Art.[107]

Roman wares, because of their far-reaching distribution and significance to trade patterns, have been repeatedly investigated by chemical analysis. Indeed, the first paper from the Oxford laboratory dealt with neutron activation of Samian ware[21] or 'Terra Sigillata', as did several early papers of the Brookhaven group.[18, 19] More recently a well-provenienced group of sherds has been analysed by Banterla *et al.*;[108] unfortunately the authors reported only peak ratios, precluding easy comparison with other data. Picon's group at Lyon have investigated compositional patterns of Terra Sigillata found at different provincial workshops in France, comparing these with each other and with wares from the original factory at Arezzo.[28, 62, 109] In their most recent paper[62] they set forth a statistical approach that is essentially a Mahalanobis-distance procedure for the assignment of probabilities of group membership. Their analyses are both by emission spectrography and by neutron activation.

Mesoamerican ceramic studies have been a speciality of the Brookhaven group ever since the very early work of Sayre *et al.* on the Mayan 'Fine Orange' wares.[19] The evolution of studies in this area will be traced here, in some detail, as an example of a pattern of successful interaction between archaeologists and physical scientists, and how the feedback can operate to stimulate innovation and further research. Interest in the fine-paste ceramic tradition of the terminal Classic phase of the Mayan civilization has been strong among archaeologists for many years in that the appearance of fine-paste (*i.e.* non-tempered) pottery appeared to parallel the rather sudden demise of the Classic Mayan cities: if the great turning point were reached as a result of intrusion of alien peoples, in, for example, the same way those bearing Toltec attributes arrived at Chichen Itza, and those aliens were identified as bearers of the fine-paste tradition, then the study of this pottery, its types and varieties, its timing, and, by analysis, its grouping and possible clay sources, assumes a considerable importance.

In the first major effort in this study Sayre and Chan[44] analysed Fine Orange from a number of sites along the Usumacinta River in Guatemala and at Chiapas, Mexico, from the Yucatan, and from Belize the wares covering the three main stylistic types which the archaeologists have designated X, Y, and Z. In his commentary appended to the paper, J. Sabloff, the archaeologist who had supplied the sherds, noted that the results, rather than yielding final answers, had in fact provided a stimulus for much additional research, and the formulation of new hypotheses for testing. The X, Y, and Z sherds were found to have broad similarity within and between groups, but also showed subtle differences that warranted further research.

Following this publication, contact was established with Professors R. Rands (Southern Illinois University) who had long been active in ceramic studies in the

[107] M. Jenkins, L. Van Zelst, and P. Meyers, private communication.
[108] G. Banterla, A. Stenico, M. Terrani, and S. Villani, *Archaeometry*, 1973, **15**, 209.
[109] M. Picon, M. Vichy, and E. Meille, *Archaeometry*, 1971, **13**, 191.

Maya Lowlands, M. Coe (Yale), and J. Paddock (Institute of Oaxacan Studies), all of whom contributed additional fine-paste material relevant to the original investigation. Through the interest and active collaboration of Rands, Bishop, and others, the study soon broadened to include tempered pottery and, most significantly, in a completely new direction, the combined petrographic (binocular microscope and thin-section) and neutron activation analyses of the same corpus of sherds. The sherds were carefully chosen from tight archaeological stratigraphic work in and around the Mayan centre of Palenque: nearly one thousand have now been analysed by both techniques and the data subjected to computer processing to establish a taxonomy based on both petrographic and chemical variables. The results, which are beginning to be published,[110] indicate that in general both petrographic thin-section and chemical variables lead to similar groups, but that the combined approach is more powerful than either alone.

A separate Brookhaven project has involved a collaboration with the University of Rochester Teotihuacan Project: here Dr. E. Rattray provided a number of sherds pertaining to the overall problem of trace and external relations of the greatest of Mesoamerican cities, Teotihuacan. The first report of this work noted that much of the Oaxacan-style pottery found at Teotihuacan was in fact locally made.[51] The same was found to be true of a series of Olmec figurines found at Tlatilco, near Mexico City, which, it had been thought, might have been imported from the Vera Cruz gulf coast. More recently, additional Olmec figurines of the same type, but from sites in Morelos, Mexico, have been analysed by Grennes and Wilson:[111] evidence of diffusion of these interesting, perhaps religiously linked, statuettes is seen.

Another important ware associated with Teotihuacan was the technologically advanced and rather unique 'Thin Orange': so close is the association that the presence of sherds of thin orange in a particular stratigraphic level is usually taken as evidence of Teotihuacan contact or influence. Here the Brookhaven group, in collaboration with Arq. R. Abascal of the Mexican Institute of Anthropology and History, has analysed sherds from a number of Mesoamerican sites.[112] The results suggest a common origin for most of the material, some of which travelled more than a thousand miles, with some examples of local imitation also in evidence.

The work on the Cholula polychrome pottery mentioned earlier has had an interesting sequel: Dr. J. Olin of the Smithsonian Institution, in a collaborative investigation of the origins and trade patterns of pottery of the Spanish Colonial period (fifteenth to eighteenth centuries) in the Americas, has made use of some of the polychrome data in her work on clay sources: the manufacturing centres for polychrome, which may have been located in Aztec times in what is now the Mexican state of Puebla, are thought to have continued to make pottery after the Spanish conquest, and so might have used the same clays. This work is still in progress,[113]

[110] R. L. Rands, P. H. Benson, R. L. Bishop, P. Y. Chen, G. Harbottle, B. C. Rands, and E. V. Sayre, 'Proceedings XLI Congreso Internacional de Americanistas, Mexico City, September, 1974', in press.
[111] R. A. Grennes and R. Wilson, paper presented at the 1975 Symposium on Archaeometry and Archaeological Prospection, Research Laboratory of Archaeology, Oxford University.
[112] R. Abascal Macias, Thesis for the title of Archaeologist, National School of Anthropology and History, and Master of Anthropological Science, University of Mexico, 1974.
[113] J. S. Olin and E. V. Sayre, *Bull. Amer. Inst. Conservation Historic Artistic Works*, in press.

but it illustrates the way data gathered in one investigation can be of use, unexpectedly, in another.

A second example of the same sort of interaction occurred in the Lubaantun Project. Lubaantun, a classic Maya site located in Belize, and its ceramics, as well as those of the surrounding area, have been studied by Professor N. Hammond, of Cambridge and Bradford Universities. A number of sherds from this area were analysed at Brookhaven. The data have been processed by three radically differing clustering procedures, a conventional average-linkage (AGCLUS, see above) technique, a new probabilistic, iterative algorithm devised by T. Gazzard[114] (I.C.I., Wilmslow, Chesire), and multidimensional scaling, for comparison. One of the Lubaantun sherds was of a distinctive style group called 'Pabellon modelled-carved' of which we had already analysed a substantial number, coming from Altar de Sacrificios and Seibal, far to the west. The Lubaantun sherd indeed closely matched the Altar–Seibal material, and greatly differed from the presumably local Lubaantun ceramics.

Archaeological ceramics in other areas are also under investigation. We have already mentioned the study of English and American pottery of the American Colonial period.[99] At the 1975 Oxford Symposium, Hunter and Warren reported on early medieval calcite-gritted wares in southern England, the so-called St. Neot's ware. Aspinall and Slater[115] have also analysed medieval British ceramics, but from a large number of kiln sites. As work of this sort accumulates, in a particular geographic area, the need for intercomparability becomes very evident.

Ancient Glass and Faience—The work of Sayre[23] involving the application of Ge–Li counters to the activation analysis of ancient glass is mentioned above. Earlier reports[22] have explored the compositional categories of ancient glasses, in particular, and detailed the changes of glass technology through history. Quite recently, Sayre and Smith have published an analytical study confined to ancient Egyptian glass.[116] Anyone interested in this field should, of course, consult the classic work of Caley[1] as well.

The problem of provenience-clustering in faience beads has already been mentioned above: the beads analysed by spectrography by Stone and Thomas,[84] and clustered by Newton and Renfrew,[83] have been reanalysed by Aspinall et al.[117] by neutron activation. These authors had confirmed earlier suspicions that the tin content, supplemented by scandium and caesium values, is a good discriminator of British faience beads, but they have not as yet subjected their data to a discriminant analysis. The grouping of the beads according to manufacturing sites would be valuable if it could be accomplished, in that it would give information on European Bronze-age technology and trade, a topic not easily approached by other means. Newton[118] has noted that 'ingots' or 'canes' of coloured glass may themselves have been objects of Iron-age trade, to be used in decorating neutral glass in local

---

[114] T. Gazzard. Some of the results were also reported at the 1975 Archaeometry Symposium at Oxford.
[115] A. Aspinall and D. N. Slater, *Nature*, 1968, **217**, 388.
[116] E. V. Sayre and R. W. Smith in 'Recent Advances in the Science and Technology of Materials', ed. A. Bishay, Plenum Press, New York, 1974, Vol. 3, pp. 47–70.
[117] A. Aspinall, S. E. Warren, J. G. Crummett, and R. G. Newton, *Archaeometry*, 1972, **14**, 27.
[118] R. G. Newton, *Archaeometry*, 1971, **13**, 11.

centres, and has pointed out the role of analysis in pursuit of evidence for this suggestion.

Finally, mention should be made of the work of Davison[119] on the neutron activation and X-ray fluorescence analyses of African trade beads, and their exchange network.

Obsidian.—Obsidian, a volcanic glass similar in composition to granite, was used throughout the whole primitive world for the production of blades, scrapers and projectile points, the conchoidal fracture and ease of flaking allowing the manufacture of very superior, sharp-edged tools. Occasionally it is found in works of art as well, e.g. the superbly executed Aztec effigies and jewelry. Its great value to primitive society, coupled with its rare occurrence in Nature, dictate that it would have been an extremely important item of commerce. It seems possible that obsidian played a crucial role in the growth of the greatest of Mesoamerican city-states, Teotihuacan, already noted for its ceramics. I quote Millon: 'It is indisputable that the population of Teotihuacan was impressively large for an early city . . . we have found more than five hundred craft workshop areas . . . the vast majority are obsidian workshops . . . . Did the growth potential represented by the expansion of the craft of obsidian working play a significant role in the rise of Teotihuacan as a city?'[120] The obsidian industry likewise was of importance in the rise of the Olmec civilization.[121] Cann et al.[122] have discussed the European and Asian obsidian trade in Neolithic times, and give many references.

The analysis of obsidian sources and artifacts easily yields results of real archaeological value because of the fortunate circumstance that obsidian, when molten in the period of volcanic activity giving rise to its formation, seems to have been 'well-stirred', i.e. the sources are often relatively internally homogeneous, and, very fortunately, differ significantly from one to the next.[121, 122] Obsidian has been analysed by emission spectrography,[122] X-ray fluorescence,[31] and neutron activation:[32, 123] in the realm of data-handling, the greatest need now seems to be to make the large amount of obsidian data more readily available and intercomparable. We have already mentioned the occasional appearance of highly correlated compositions[77] and the use of Mahalanobis distance in relating obsidian artifacts to sources.[86] Obsidian analysis has become a standard archaeological laboratory tool, and deserves a review on its own merit: it has already contributed heavily to our knowledge of prehistoric trade patterns.

Precious Metals and Coins.—Nuclear activation presents the archaeologist, numismatist, conservator, or art historian with a most important technique in that pre-

[119] C. C. Davison, Ph.D. Thesis, University of California, 1972; see Lawrence Berkeley Laboratory Report 1240, 1972.
[120] R. Millon, in 'Urbanization at Teotihuacán, Mexico', ed. R. Millon, Vol. I, Part 1, University of Texas Press, Austin, 1973; *Science*, 1970, **170**, 1077.
[121] R. H. Cobean, M. D. Coe, E. A. Perry, jun., K. K. Turekian, and D. P. Kharkar, *Science*, 1971, **174**, 666.
[122] J. R. Cann, J. E. Dixon, and C. Renfrew, in 'Science in Archaeology,' ed. D. Brothwell and E. Higgs, Thames and Hudson, 2nd Edn., 1969, Chapter 51, pp. 578—591; *Scientific American*, 1968, **218** (3), 38.
[123] Many references to obsidian analysis will be found in the AATA abstracts (ref. 17) and *Archaeometry*. See also G. E. Coote, N. E. Whitehead, and G. J. McCallum, *J. Radioanalyt. Chem.*, 1972, **12**, 491.

cious, irreplaceable objects may be analysed non-destructively, or by the removal of minute samples, in cases where the grosser demands of conventional analytical sampling could not be met. Coin analysis by activation has been widely practised: much of the work related to historical periods, which probably ought not to be included under 'Archaeology'. And yet, there is a lesson to be learned: even in coinage, where mixing, remelting, and recasting presumably occurred, the metal often shows trace-element patterns which may be indicative of the original mining areas. Among the many studies of ancient coins we may mention those of Kraay and Emeleus,[124] Das and Zonderhuis,[125] Bluyssen and Smith,[126] Gordus,[37,127] Meyers[50,128] who employed charged-particle and fast-neutron bombardment, Ravetz,[129] Wyttenbach and Hermann,[130] and Oddy,[131] the last paper containing several additional references to analysis of medieval gold coins, as well as comparative analyses. The abstracts published by the International Institute for Conservation[17] are a rich mine of references on coin analysis, by all techniques.

There has been a substantial interest in the neutron activation analysis of silver objects of the Sasanian period (A.D. 224—651). Lechtman[132] reports analyses by Sayre on a silver rhyton in the form of a horse, from the Cleveland Museum of Art. Here, the purpose was to investigate the technical means by which the gilding was applied to the horse. But other research, on Sasanian coins[127] and museum objects,[133] has concentrated more on the trace element patterns as a provenience guide, and for autdentication. An interesting technical development is the perfection of the 'streak' method: a minute (50—100 µg) sample of the silver object is rubbed off onto a clean quartz tube. After bombardment, the silver, copper, and gold are measured by INAA. Bombarded drilled samples are dissolved and processed to remove the high silver, copper, and gold activities. The filtrate then contains the remaining trace elements, which are determined by Ge–Li counters.[38,127] The streak method was used to analyse a proto-Elamite silver figurine from the Metropolitan Museum: here an important question was the nature of the solder employed.[134]

Some gold objects of great archaeological interest have also been analysed by neutron activation, particularly to get evidence of ancient technology (*e.g.* the Moulsford Torc—an Irish gold neck-ring of the Middle Bronze Age, weighing nearly a pound[135]). Similarly, an electrum diadem from Egypt (1650 B.C.)[136] and gold figurines from Peru[137] have been examined. Finally, it ought to be mentioned that activation analysis of ancient gold objects has also been carried out with

---

[124] C. M. Kraay and V. M. Emeleus, *Archaeometry*, 1959, **2**, 1.
[125] H. A. Das and J. Zonderhuis, *Archaeometry*, 1964, **7**, 90.
[126] H. Bluyssen and Ph. B. Smith, *Archaeometry*, 1962, **5**, 113.
[127] A. A. Gordus, in ref. 31, pp. 145—155; A. A. Gordus and J. P. Gordus, in ref. 38., pp. 124—147
[128] P Meyers, Ph.D. Thesis, University of Amsterdam, 1968.
[129] A. Ravetz, *Archaeometry*, 1963, **6**, 46.
[130] A. Wyttenbach and H. Hermann, *Archaeometry*, 1966, **9**, 139.
[131] W. A. Oddy, *Archaeometry*, 1972, **14**, 109.
[132] H. N. Lechtman, in ref. 31, pp. 2—30.
[133] P. Meyers, L. Van Zelst, and E. V. Sayre, *J. Radioanalyt. Chem.*, 1973, **16**, 67.
[134] K. C. Lefferts, *Metropolitan Museum J.*, 1970, **3**, 15—24.
[135] E. T. Hall and G. Roberts, *Archaeometry*, 1962, **5**, 28.
[136] K. C. Lefferts, *Bull. Metropolitan Museum of Art*, 1969, **28**, 61.
[137] H. N. Lechtman, L. A. Parsons, and W. J. Young, in 'Studies in Pre-Columbian Art and Archaeology No. 16', Dumbarton Oaks Trustees for Harvard University, Washington, D.C., 1975, pp. 7—46.

protons, by Barrandon et al.[138] The use of very energetic protons, or fast neutrons, or high-energy photons[50, 128] could avoid difficulties with self-shielding which would surely arise with the irradiation of massive gold objects with thermal neutrons.

Turquoise, Jade, and Amber.—These three chemically very dissimilar materials are lumped together in that all were items of luxury trade in ancient times (as they are today), hence, in all likelihood, subject to quite different economic forces from those governing utilitarian materials like obsidian, pottery, and flint. The turquoise was mined and treasured both in the Old and New Worlds: the excellent monograph of Pogue[139] is encyclopaedic on the subject. Shakespeare has Shylock say that he would not have traded his turquoise ring 'for a whole wilderness of monkeys'. In Mesoamerica turquoise had enormous importance: its use was limited to priests, kings, and nobility. As Cortez approached Mexico City, the Emperor Montezuma, thinking the god Quetzalcoatl had returned, sent him a mask of turquoise mosaic, to be seen today at the British Museum.[140]

Although the stone was much valued in Mesoamerica, there are few mines there, and for many years archaeologists had speculated that a 'turquoise road' might have extended from central Mexico up to the southwestern United States, where the mining areas in Arizona, New Mexico, Colorado, and Nevada show ample evidence of aboriginal working. The joint project between the Department of Anthropology at the State University of New York (Stony Brook) and Brookhaven Laboratory has for the past three years been analysing turquoise and other green and blue stones (malachite, azurite, chrysocolla *etc.*) used culturally in pre-Columbian America, concentrating first on mining source areas. Fifty or more, including some in Mexico, have been sampled, and the research is turning now to the analysis of artifacts; mosaic blanks, beads, and amulets. At the University of Arizona, Sigleo[42] has also carried out neutron activation analysis of turquoise: beads from the Snaketown site were traced to a mine in south-eastern California, on the basis of Co, Cr, Eu, Sb, Sc, and Ta contents.

Jade was likewise a luxury item in Mesoamerica: the superb Olmec and Mayan carvings and masks convey a powerful sense of artistic taste and religious fervor.[141] Again, there are but few sources known:[142] some of these are now in the process of being analysed, and the initial results appear promising.[143]

In the case of amber, the best technique of characterization is that devised by Beck et al.[144] based upon computer classification of the i.r. spectra. However, there is one paper by Das[145] on activation analysis of archaeological amber: Baltic and

---

[138] J. N. Barrandon, J. L. De Brun, P. Benaben, and C. Rouxel, *Ann. du Lab. de Recherche des Museés de France*, 1973, 59—63.
[139] J. E. Pogue, 'Turquois', Memoirs of the National Academy of Sciences XII, Part II, 1915, Reprinted by Rio Grande Press, Glorieta, New Mexico, 1974.
[140] E. Carmichael, 'Turquoise Mosaics from Mexico', Trustees of the British Museum, London, 1970.
[141] R. L. Rands and M. D. Coe, in 'Handbook of Middle American Indians', ed. G. R. Willey, Part 2, Vol. 3, University of Texas Press, Austin, 1965, articles 21 and 29 respectively.
[142] W. F. Foshag, *Smithsonian Miscell. Collections*, 1957, **135**, 5, Publication 4307, Washington D.C.
[143] N. Hammond, personal communication.
[144] C. W. Beck, A. B. Adams, G. C. Southard, and C. Fellows, in ref. 31, pp. 235—243.
[145] H. A. Das, *Radiochem. Radioanalyt. Letters*, 1969, **1**, 289.

Sicilian ambers were found to differ in their sodium and gold contents. This would seem to be a promising area for additional research.

Other Stones of Archaeological Interest: Sandstone, Steatite, Sanukite, Marble, and Flint.—This section deals with a miscellany of stones, and, in most cases, limited investigation. But it is hoped that it may serve to stimulate archaeologists to try provenience-oriented analytical studies of these or others, *e.g.* the specular haematite or cinnabar associated with the Olmec civilization.

In an interesting paper, Heizer *et al.*[146] have investigated the source of the quartzose sandstone, or 'quartzite' of the two 720-ton 'Colossi' of Memnon at Thebes in Egypt: the more northerly of these was the famous 'singing' Memnon of antiquity. This statue was repaired by the Emperor Septimius Severus in Roman times, and Heizer *et al.* also analysed the stone used in the repair. The results suggest that the quarries at Gebel el Ahmar, near Cairo, 676 km down the Nile, were the source of the stone for the original statues, but that the stone used in the Septimius Severus repair came from the nearer quarries at Aswan and Edfu, upstream.

Steatite, or 'soapstone', was widely used in both Asia and America: in both areas the mining and trade in soapstone artifacts extends back for many millenia. Here, again, sources tend to be few, and trade-routes correspondingly long.[147] At Brookhaven, an extensive project related to the steatite trade emanating from Tepe Yahya, a third-millenium B.C. site in south-eastern Iran, was undertaken by Kohl.[148] It was found that neutron activation was unsuccessful, since the compositional spread in samples from a given mining source was extremely broad. It was also found, by X-ray diffraction, that the material in question was not steatite at all, but chlorite. However, in the end Kohl was able to characterize the material by measuring the relative intensities of the first five basal plane reflections, and performing cluster analysis on ratios of intensities.

With American Indian soapstone artifacts and sources, Luckenbach *et al.* were able to apply neutron activation in an innovative way, that proved to be completely successful.[149] They found, as in the case of the Tepe Yahya chlorite, great scatter in the concentrations of a particular element in repetitive samples from the same quarry. But they discovered that the pattern of concentrations *within* the rare-earth group tended to be much more reproducible, and furthermore, that this pattern varied sharply from quarry to quarry. In this fashion they were able to establish the source of soapstone used in artifacts: their technique is surely worth trying in other systems as well.

Sanukite is a black andesitic rock, often conchoidally fractured, from which stone implements were fashioned in the period from the first century B.C. to the first century A.D. in Japan. Although X-ray fluorescence was employed in an element determination by Higashimura and Warashina,[150] this work is mentioned as another

[146] R. F. Heizer, F. Stross, T. R. Hester, A. Albee, I. Perlman, F. Asaro, and H. Bowman, *Science*, 1973, **182**, 1219.
[147] E. Porada, *Artibus Asiae*, 1971, **33**, 323.
[148] P. L. Kohl, Ph.D. Thesis, Department of Anthropology, Harvard University, Cambridge, Massachussetts, 1974.
[149] A. H. Luckenbach, C. G. Holland, and R. O. Allen, *Science*, 1975, **187**, 57; *Archaeometry*, 1975, **17**, 69.
[150] T. Higashimura and T. Warashina, Research Reactor Institute, Kyoto University, Japan, personal communication.

example of a stone that could be successfully traced to its source through chemical composition. The authors have employed Mahalanobis distance in their data-handling.

Marble from quarries which fed the ateliers of the sculptors of classic Greece has been characterized by petrological methods[151] and more recently by measurements of the $^{13}C:^{12}C$ and $^{18}O:^{16}O$ isotopic ratios.[152,153] Although the isotopic ratio method gives good provenience separation, it does require rather specialized equipment and personnel. It would therefore be of interest to see if nuclear activation, first tried by Rybach and Nissen,[154] might also be applicable. The X-ray fluorescence analyses of Conforto *et al.*, which dealt with major rather than trace elements, did not, on the whole, produce good separations,[155] but an examination of the whole range of trace elements might lead to better discrimination.

The importance of flint in European prehistory is attested by Sieveking *et al.*, 'Artefacts made from flint are the commonest and indeed almost the only surviving relics in Western Europe for by far the greater part of Man's existence'.[156] Their paper, which lists analyses carried out by atomic absorption, also contains an excellent discussion of the geology of flint: the authors processed their data by discriminant analysis, appropriate to their combination of limited number of source areas, large number of artifacts. Bruin *et al.*[157] have made use of 'pattern recognition' in the numerical taxonomy of flints. These scientists employed neutron activation and measured 14 elements, and they remark that the more limited number of elements used by Sieveking in his earlier work[158] may lead to unreliable assignments. One of the more interesting items published by Bruin *et al.* is a table of 10 replicate analyses from each of two single flint pieces: the great range of values is reminiscent of the Iranian chlorite (see above).

Aspinall and Feather have also analysed flint by NAA, for a total of 15 elements.[36] Again, they noted large variations for given elements within a particular source, or indeed, within one single piece. They found that flint contains correlated elements, especially the rare earths, and concluded that although some single elements might be useful in characterization, ratios of elements might be better. One must conclude that much additional work remains to be done with flint, but that analysis gives a distinct promise of source characterization.

Steel, Bronze, Copper, and Lead.—The archaeologist may be interested in metals like bronze and copper from several different viewpoints—in the economic sphere, as coins; in the technology and trade of artifacts like tools, pins *etc.*; and in the artistic use of the metals in figurines and jewelry. Great numbers of analyses of ancient metal objects have been made by emission spectrography, wet chemistry, and X-ray

[151] C. Renfrew and J. Springer Peacy, *J. Brit. School of Archaeology at Athens*, 1968, **63**, 45.
[152] H. Craig and V. Craig, *Science*, 1972, **176**, 401.
[153] L. Manfra, U. Masi, and B. Turi, *Archaeometry*, 1975, **17**, 215.
[154] L. Rybach and H.-U. Nissen, in 'Radiochemical Methods of Analysis', International Atomic Energy Agency, Vienna, 1965, Vol. I, pp. 105—117.
[155] L. Conforto, M. Felici, D. Monna, L. Serva, and A. Taddeucci, *Archaeometry*, 1975, **17**, 201.
[156] G. de G. Sieveking, P. Bush, J. Ferguson, P. T. Craddock, M. J. Hughes, and M. R. Crowell, *Archaeometry*, 1972, **14**, 151.
[157] M. De Bruin, P. J. M. Korthoven, C. C. Bakels, and F. C. A. Groen, *Archaeometry*, 1972, **14**, 55.
[158] G. de G. Sieveking, P. Craddock, M. J. Hughes, P. Bush, and J. Ferguson, *Nature*, 1970, **228**, 251.

fluorescence but attention will be focussed on those made by nuclear techniques. Meyers[50, 128] has reported on the analysis of coins of orichalcum, a Roman copper–zinc alloy, and some Bronze Age objects from Luristan, Egypt, and Brittany, using fast neutrons. We have already mentioned the interesting round-robin analysis project sponsored by the Freer Gallery:[48] the materials were a Luristan and a Shang Dynasty bronze, and several of the participating laboratories employed neutron activation.

In antiquity, copper was smelted from a variety of ores and also found in the native state. The trade patterns in native copper in North America are particularly interesting in that a single large source in the Lake Superior region was heavily exploited:[159] in South America, the Incas generally smelted copper ores rather than use native copper.[160] The Argonne group headed by Friedman have specialized in the analysis of native copper, copper smelted from different kinds of ores, and copper artifacts.[61, 161, 162] They have devised a mathematical system, given the analysis of an artifact, for predicting the type of ore from which the copper was won. Unfortunately, they have published only a very small selection of their actual analytical data,[160] preferring rather to publish statistical summaries.[162] In their study of copper objects from the Moche culture, Friedman *et al.* found a significant silver–gold correlation, which may prove to be useful in taxonomy.[160]

Although steel (like most other metals from antiquity) has usually been studied by metallography, Voigt and Abu-Samra[163] have measured the carbon in a Damascus blade by inducing the reaction $^{12}C(\gamma,n)^{11}C$ with high-energy photons and other elements by neutron activation.

Roman lead has been analysed by Wyttenbach and Schubiger,[39] who rightly noted that '[analytical] information on lead is rather scarce. This is all the more surprising as the Romans are famous for their extensive production and copious use of lead, especially for water systems. Due to the high resistance of lead to corrosion, finds of artifacts of lead are plentiful'. A happy technical advantage is that the lead matrix itself gives very little radioactivity upon neutron activation, allowing one easily to 'see' the trace elements. Although Wyttenbach and Schubiger were able to demonstrate some features of Roman pipe construction methods, they did not find clustering of samples based on composition. This they felt was probably due to remelting, or mixing of raw lead from different origins.

5 Conclusions

It is hoped that this broad and necessarily incomplete survey of the applications of activation analysis to archaeology will serve to attract the attention of more physical scientists to the broader field of archaeometry. For here is a rapidly developing interdisciplinary field where human ingenuity can still compete on equal terms with the great machines of modern physical science, and where the humanist

---

[159] C. C. Patterson, *Amer. Antiq.*, 1971, **36**, 286.
[160] A. M. Friedman, E. Olsen, and J. B. Bird, *Amer. Antiq.*, 1972, **37**, 254.
[161] A. M. Friedman, M. Conway, M. Kastner, J. Milsted, D. Metta, P. R. Fields, and E. Olsen, *Science*, 1966, **152**, 1504.
[162] P. R. Fields, J. Milsted, E. Henrickson, and R. Ramette, in ref. 31, pp. 131—143.
[163] A. F. Voigt and A. Abu-Samra, 'Proceedings 1965 International Conference on Modern Trends in Activation Analysis', Texas A. and M. University, College Station, Texas, 1965, pp. 22—25.

and scientist can collaborate in a very direct and natural way. In closing I could not do better than quote from the splendid, but modestly headed, 'Post-Symposium Note: Science in the Service of History' by C. S. Smith, who was speaking of exactly this humanist–scientist interaction.[164] 'The lack of concern with technology has produced a common view of history that is utterly unrealistic and one that provides a singularly poor basis for understanding the predicament of the present world. Scientists should themselves be more interested than any other group in correcting the perspective. For they are the direct inheritors of the old tradition. It is particularly appropriate for them to do so at the present time because they alone are familiar with the recently developed analytical procedures that are necessary. Let scientists, therefore, ask and answer their own historical questions as well as help to answer those presently posed by the archaeologist and general historian. In so doing not only will they come to understand their own profession better but they will also extend the range of what is called history. Historians have always exploited the contents of libraries, and they are beginning to make more professional use of museums. With initial help from scientists they will come to use the laboratory also, and the more they do so, the more they will be able to understand the origin of ideas and the forces for cultural change.'

The author acknowledges with pleasure a deep debt of gratitude to his colleagues E. V. Sayre, A. Bieber, jun., R. W. Bishop, P. Meyers, and L. van Zelst for many stimulating discussions, for continuing instruction and assistance. All have helped to make the Brookhaven Art and Archaeometry group more than the sum of its parts. I also thank R.W. Dodson for invaluable help with the historical record, and E. V. Sayre for the preparation of Figures 1—5. The present paper was written under the auspices of the U.S. Energy Research and Development Administration.

[164] C. S. Smith, in ref. 31, pp. 53—54.

3
Preparation of Radiopharmaceuticals and Labelled Compounds using Short-lived Radionuclides

BY D. J. SILVESTER

1 Introduction

This subject has not been reviewed in the purely chemical literature before, and it may therefore be as well to begin with a few words of general introduction.

A radiopharmaceutical is a medicinal product, generally used in the investigation of human disease, which contains a radionuclide as an integral part of the main ingredient.[1] Labelled compounds thus include radiopharmaceuticals, but may be used in a much wider range of applications. However, because of this Reporter's interests, and because of the further restriction of this review to short-lived radionuclides, explained below, most of the compounds discussed have applications in medicine or closely related disciplines.

The principles of the radioactive technique were established by Hevesey[2,3] some 60 years ago, and since the first discovery of artificial radionuclides by Joliot and Curie[4] the technique has undergone continuous development. Much was, and still can be, accomplished by using radionuclides in simple chemical forms, *e.g.* radioiodide in mapping the distribution of functioning thyroid tissue, or radiokrypton gas in measuring regional lung ventilation or cardiac blood flow. However, many investigations nowadays call for the use of relatively complex radioactive compounds, *e.g.* ^{11}C-labelled catecholamines, or ^{123}I-labelled proteins, the synthesis of which can raise many new problems to intrigue the small but growing number of chemists working on them.

The description 'short-lived' in this context needs definition. Although radionuclides with half-lives as short as 2 min (^{15}O) and even 13 s (^{81m}Kr) are currently in frequent use, 10 min ^{13}N is the shortest-lived radionuclide yet to have been employed in true synthetic procedures.[5] The upper limit of 'short' is more difficult to decide, but the convention seems to be established that half-lives greater than about 100 h are 'long'. Thus, compounds incorporating ^{131}I(8 d), ^{3}H(12.3 y), and ^{14}C(5730 y) are amongst those excluded from this review.

It may be asked what justification there is for separate consideration of 'short-lived' radionuclides. In practical application, they can offer several advantages over

---
[1] 'Guidelines for the Preparation of Radiopharmaceuticals in Hospitals', Special Report No. 11, British Institute of Radiology, London, 1975.
[2] G. Hevesy, *Z. anorg. Chem*, 1913, **82**, 323.
[3] G. Hevesy, *Biochem. J.*, 1923, **17**, 439.
[4] F. Joliot and I. Curie, *Nature*, 1934, **133**, 201.
[5] A. P. Wolf, D. R. Christman, J. S. Fowler, and R. M. Lambrecht, in 'Radiopharmaceuticals and Labelled Compounds', I.A.E.A., Vienna, 1973, Vol. 1, p. 345.

longer lived radionuclides, including the obvious one that repetitive measurements can be made without a general increase in background activity levels. In clinical use, the radiation dose to patients can be much reduced, or, for a given radiation dose, higher activities can be administrated, resulting in more accurate data being obtained. Furthermore, for some elements—including several essential constituents of living matter, *e.g.* carbon, nitrogen, and oxygen—the only radionuclides which can be readily detected and measured *in vivo* are short-lived ones (^{11}C, ^{13}N, and ^{15}O). From a chemist's point of view, however, the preparation of compounds incorporating such radionuclides calls for the development of rapid synthetic and analytical techniques which are frequently quite different from the conventional ones which can be employed with long-lived radionuclides.

The starting point for this review was the I.A.E.A. Symposium on Radiopharmaceuticals and Labelled Compounds at Copenhagen in 1973, the proceedings of which were published[6] and serve as a general source of information on earlier work. Recently, the proceedings of another Symposium on this subject, held at Atlanta in 1974, have been published[7] and contain many useful review articles, as does a special issue of *Seminars in Nuclear Medicine*.[8] A standard earlier reference work is another I.A.E.A. publication.[9] The applications of radiopharmaceuticals have been frequently reviewed, and the published proceedings of two further meetings[10,11] are particularly worth noting in this regard.

The basic facilities and working procedures required for the safe preparation of radiopharmaceuticals in hospitals were the subject of a seminar held at the Royal Marsden Hospital, Sutton, in October 1974. Following this, a working party was set up with the task of preparing some guidelines on the subject and these have now been published.[1] Recipes for the preparation of particular radiopharmaceuticals are not included, but the general principles (which are just as applicable to laboratories outside hospitals) are comprehensively reviewed.

2 Sources of Short-lived Radionuclides

Short-lived radionuclides may be obtained directly as a result of nuclear reactions occurring in reactors or in charged-particle accelerators, or indirectly from the decay of a longer-lived parent radionuclide in a so-called generator system.

The essential differences between accelerator and reactor products have been summarized.[12] Although there are naturally many exceptions, as a general rule reactor-produced radionuclides are neutron-rich and decay by β^- emission, whilst charged-particle-induced reactions much more often produce neutron-deficient radionuclides, which decay by β^+ emission or by electron capture. For many purposes, the physical properties of neutron-deficient radionuclides are advantageous.

[6] 'Radiopharmaceuticals and Labelled Compounds', I.A.E.A., Vienna, 1973, Vols. 1 and 2.
[7] 'Radiopharmaceuticals', ed. G. Subramanian, B. A. Rhodes, J. F. Cooper, and V. J. Sodd, The Society of Nuclear Medicine, New York, 1975.
[8] *Seminars in Nuclear Medicine*, 1974, **4**, No. 3.
[9] 'Radioisotope Production and Quality Control', Technical Reports Series No. 128, I.A.E.A., Vienna, 1971.
[10] 'New Techniques in Tumour Localization and Radioimmunoassay', ed. M. N. Croll, L. W. Brady, T. Honda, and R.J. Wallner, Wiley, New York, 1974.
[11] 'Dynamic Studies with Radioisotopes in Medicine 1974', I.A.E.A., Vienna, 1975, Vols. 1 and 2.
[12] R. S. Tilbury and J. S. Laughlin, ref. 8, p. 245.

Preparation of Radiopharmaceuticals and Labelled Compounds

Furthermore, some very useful radionuclides (*e.g.* ^{11}C, ^{13}N, ^{15}O) can only be made satisfactorily by charged-particle induced reactions.

It is worth noting further that, whereas the products of (n,γ) reactions which predominate in reactors are usually of low specific activity (apart from the relatively rare occasions when advantage can be taken of the effects of nuclear recoil), many charged-particle induced reactions [*e.g.* (p,*x*n), (d,*x*n), (d,α), (α,*x*n), (α,2*x*n) *etc.*] and all generator systems, yield products which are not isotopic with the target element or parent radionuclide, and so can be isolated in very high specific activity, or even 'carrier-free'.

Nuclear Reactors.—The basic principles governing the production of radionuclides in nuclear reactors are described in an article by Poggenburg,[13] who points out that several factors, including typically large thermal neutron fluxes and reaction cross-sections, contribute to giving reactors a much larger capacity for radionuclide production than charged-particle accelerators. The article includes a reference table to 40 reactor-produced radionuclides, of which 15 are 'short-lived' by our definition. These are products of (n,γ) reactions (^{24}Na, ^{42}K, ^{47}Sc, ^{69m}Zn, ^{97}Ru, ^{113m}In, ^{157}Dy, ^{171}Er, ^{195m}Pt, ^{198}Au, ^{199}Au, and ^{197}Hg) of (n,p) reactions (^{43}K, ^{67}Cu, and ^{69m}Zn) or of the (n,fission) reaction (^{99}Mo). Some discussion of practical considerations, including the selection of target compounds and isotopic enrichment, is included.

Charged-particle Accelerators.—Work published before 1973 on the preparation of medically useful radionuclides by accelerators has been reviewed previously.[12,14]

Although many kinds of accelerators have been used, ranging from 14 MeV neutron generators[15,16] and small van de Graaf machines[17] to the 200 MeV proton linear accelerators at Brookhaven National Laboratory,[18] by far the most widely used machine for radionuclide production is the cyclotron,[19] especially since advances in engineering have resulted in the development of 'compact' machines[20] which are relatively inexpensive to build and to operate. Whilst many larger cyclotrons, originally built for work in nuclear physics, are now used for radionuclide production, compact machines have been installed in many centres[12,14] to serve just this purpose.

The most desirable physical characteristics of such machines, the ancillary facilities that are needed, and the special problems in radiopharmaceutical production attributable to the use of a cyclotron, rather than a reactor or generator, as the radionuclide source, have been discussed in two papers.[19,21] The first is a report prepared by an I.A.E.A. Consultants' panel, and the second is based on some 15 years experience with the M.R.C. cyclotron at Hammersmith Hospital.

[13] J. K. Poggenburg, ref. 8, p. 229.
[14] D. J. Silvester, ref. 6, Vol. 1, p. 197.
[15] Z. B. Alfassi and A.P. Kushelevsky, *Radiochem. Radioanalyt. Letters*, 1975, **20**, 347.
[16] Z. B. Alfassi and A. P. Kushelevsky, *Radiochem. Radioanalyt. Letters*, 1975, **21**, 87.
[17] R. D. Moore and H. Troughton, *Photosynthetica (Prague)*, 1973, **7**, 271.
[18] P. Richards, E. Lebowitz, and L. G. Stang, jun., ref. 6, Vol. 1, p. 325.
[19] J. L. Servian, *Internat. J. Appl. Radiation Isotopes*, 1975, **26**, 763.
[20] G. O. Hendry, in 'Cyclotrons-1972', ed. J. J. Burgerjon and A. Strathdee, American Institute of Physics, New York, 1972, p. 616.
[21] D. J. Silvester, ref. 7, Chapter 16, p. 157.

Methods used in the preparation of several radionuclides, using accelerators in Amsterdam,[22] New York,[23] Los Angeles,[24] and Tokyo,[25] have been reviewed.

Radionuclide-generator Systems.—A radionuclide generator system (formerly sometimes called a 'radioisotope cow') is one in which a short-lived daughter radionuclide is separated by some means (usually chromatography, when the process is still known as 'milking') from its longer-lived parent.

A short but authoritative review of practical generator systems by Lebowitz and Richards[26] begins by discussing the familiar equation governing such a parent–daughter relationship:

$$N_2{}^t = \frac{\lambda_1}{\lambda_2 - \lambda_1} N_1{}^0 (e^{-\lambda_1 t} - e^{-\lambda_2 t}) + N_2{}^0 e^{-\lambda_2 t} \qquad (1)$$

where the symbols have their customary meanings, subscript 1 denoting the parent and subscript 2 the daughter species).

As an example, the growth–decay curve of the ^{99}Mo–^{99m}Tc generator system (for which $\lambda_2 \approx 10\lambda_1$) is considered (Figure 1). After the daughter has been milked from the parent ($t = 0$) the daughter activity grows in from the decay of the parent until a state of transient equilibrium is reached and both appear to decay with the half-life of the parent, equation (1) reducing to:

$$N_2{}^t = \frac{\lambda_1}{\lambda_2 - \lambda_1} N_1{}^0 e^{-\lambda_1 t} \qquad (2)$$

Once separated from the parent, the daughter nuclide decays with its own characteristic half-life. The frequency with which the short-lived daughter can be milked in high yield depends upon the rate at which it grows back in equation (1). In practice the separation is made as frequently and for as long as there is sufficient activity available to carry out the desired procedures.

Lebowitz and Richards discuss five systems (those yielding 6 h ^{99m}Tc, 99 min ^{113m}In, 13 s ^{81m}Kr, 1.3 min ^{82}Rb, and 17.5 s ^{77m}Se) in some detail. Another review, by Yano,[27] tabulates the physical properties of parent and daughter nuclides for a total of 27 possible generator systems, of which 13 are reported to have been put to practical use. The nuclear reactions which may be employed to make the parent radionuclides are also listed; of 33 such reactions only nine are neutron-induced, whilst the balance are charged-particle induced.

Some aspects of radiopharmaceutical production using radionuclide generators have also been reviewed by Mitta.[28]

3 Carbon-11

Innumerable advances in all fields of study touched by organic chemistry can be

---

[22] L. Lindner, G. A. Brinkman, T. H. G. A. Suer, A. Schimmel, J. Th. Veenboer, F. H. S Karten, J. Visser, and C. J. Leurs, ref. 6, Vol. 1, p. 303.
[23] A. S. Gelbard, T. Hara, R. S. Tilbury, and J. S. Laughlin, ref. 6, Vol. 1, p. 239.
[24] N. S. MacDonald, ref. 7, Chapter 17, p. 165.
[25] Y. Murakami, F. Akiha, and O. Ezawa, ref. 6, Vol. 1, p. 257.
[26] E. Lebowitz and P. Richards, ref. 8, p. 257.
[27] Y. Yano, ref. 7, Chapter 25, p. 236.
[28] A. E. A. Mitta, *J. Radioanalyt. Chem.*, 1975, **27**, 129.

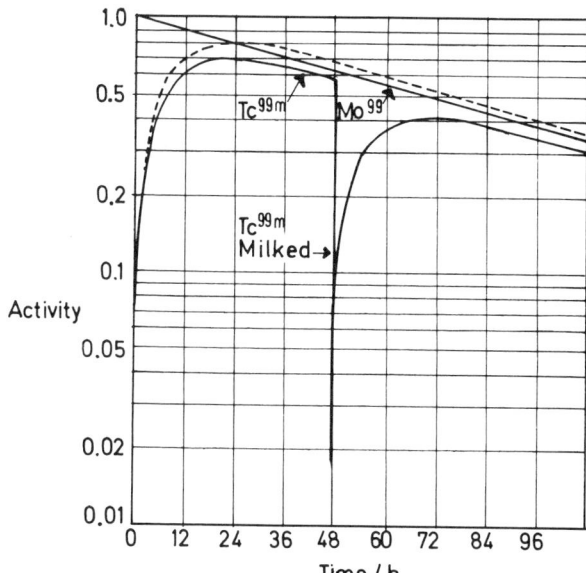

Figure 1 *Growth-decay curve of the ^{99}Mo–^{99m}Tc generator system. ^{99}Mo parent decays to ^{99m}Tc (ca. 86%) and ^{99}Tc (ca. 14%). Broken line indicates 100% parent decay through daughter.*
(Reproduced by permission from *Seminars in Nuclear Medicine*, 1974, **4**, 258)

attributed to the use of ^{14}C-labelled compounds as tracers. However, ^{14}C ($t_{\frac{1}{2}}$ = 5730 y) decays by emission only of soft β^- rays, and so cannot be detected *in vivo*. Carbon-11 ($t_{\frac{1}{2}}$ = 20.3 min), on the other hand, decays by β^+ emission, and so is readily measurable *in vivo* by detection of the penetrating 511 KeV β^+-annihilation radiation. Consequently, despite, or sometimes because of, its short half-life, many uses have been found for this radionuclide.

Methods of producing ^{11}C published prior to 1973 have been reviewed.[5,14] Recently,[29] it has been made by photospallation of oxygen, ^{16}O(γ,2p3n)^{11}C using a high-energy (2—22 GeV) electron linear accelerator. However, the nuclear reactions most commonly employed are the following.[30]

	Reaction	Q-value/MeV	Practical threshold energy/MeV
(i)	^{11}B(p,n)^{11}C	−2.76	ca. 3
(ii)	^{10}B(d,n)^{11}C	+6.47	ca. 3
(iii)	^{11}B(d,2n)^{11}C	−4.99	ca. 6
(iv)	^{14}N(p,α)^{11}C	−2.92	ca. 5

[29] J. A. DeGrazia, A. F. Rodden, J. D. Teresi, D. D. Busick, and D. R. Walz, *J. Nuclear Medicine*, 1975, **16**, 73.
[30] J. C. Clark and P. D. Buckingham, 'Short-lived Radioactive Gases for Clinical Use', Butterworths, London, 1975, Chapter 7, p. 215.

Although small but useful yields of ^{11}C have been obtained from van de Graaf accelerators[31,17] by reactions (i) and (ii), cyclotrons are undoubtedly a more prolific source and may be regarded as essential if complex syntheses are to be achieved.[32]

Either solid (usually B_2O_3) or gaseous (N_2) targets may be bombarded to make ^{11}C, but in either case the various recoil and radiation-induced chemical reactions which occur result in a high proportion of the ^{11}C atoms forming volatile compounds (*e.g.* CO, CO_2, CH_4, HCN) which can be swept out of the target vessel continuously in a stream of a suitable gas. Much detailed information on the construction and performance of several possible target and recovery systems is given in a book by Clark and Buckingham,[30] where the optimum conditions for the preparation of ^{11}CO and $^{11}CO_2$ are discussed.

$^{11}CO_2$ is an important precursor in the synthesis of several classes of compounds and so too is $H^{11}CN$. Continuous production of the latter, by 15 MeV proton bombardment of a mixture of 99% N_2 and 1% H_2, flowing at about 4 atm pressure through a heated target vessel, has been described.[33] An alternative method which has been studied and reported in detail by Christman *et al.*,[34,35] uses a similar gas mixture (94.5% N_2, 5.5% H_2) at 11 atm pressure in an unheated target, and a proton beam of 18 MeV incident energy. The primary products produced in this system are ^{11}CO, $H^{11}CN$, and a trace of $^{11}CO_2$; however, it has been established by radio-gas chromatographic analysis that, under the very high radiation doses which prevail (up to 25 eV molecule^{-1}) all such products are converted into $^{11}CH_4$ radiolytically. Additional radiolytic reactions in the target vessel result in the simultaneous formation of excess amounts of (non-radioactive) NH_3, and by passing the gas flowing out of the target through a heated platinum furnace, this quantitatively converts $^{11}CH_4$ into $H^{11}CN$. Excess NH_3 can be removed by passage through P_2O_5, and the $H^{11}CN$ collected in a cold trap, from which it may be recovered at the end of the cyclotron irradiation. With a cyclotron beam current of 30 μA, a 45 min irradiation has yielded 2 Ci of $H^{11}CN$, with no added carrier.

Syntheses with $^{11}CO_2$.—Perhaps the simplest compounds which may be synthesized from $^{11}CO_2$ are the [^{11}C]carboxylates.[36] Two general reactions have been employed in the first $^{11}CO_2$ reacts with a Grignard reagent:

$$RMgCl + {}^{11}CO_2 \rightarrow R{}^{11}CO_2MgCl \xrightarrow[\text{ii, NaHCO}_3]{\text{i, H}^+} R{}^{11}CO_2Na$$

and in the second, with an aryl-lithium compound:

$$\text{Ph-Li} + {}^{11}CO_2 \longrightarrow \text{Ph-}{}^{11}CO_2Li \xrightarrow[\text{ii, NaHCO}_3]{\text{i, H}^+} \text{Ph-}{}^{11}CO_2Na$$

[31] A. G. Perris, R. O. Lane, J. Y. Tong, and J. D. Matthews, *Internat. J. Appl. Radiation Isotopes*, 1974, **25**, 19.
[32] M. E. Phelps and B. W. Wieland, *Phys. Med. Biol.*, 1973, **18**, 284.
[33] W. G. Myers, J. F. Lamb, R. W. James, and H. S. Winchell, *Nuclear-Medicine*, 1973, **12**, 154.
[34] D. R. Christman, R. D. Finn, K. I. Karlstrom, and A. P. Wolf, *J. Nuclear Medicine*, 1973, **14**, 864.
[35] D. R. Christman, R. D. Finn, K. I. Karlstrom, and A. P. Wolf, *Internat. J. Applied Radiation Isotopes*, 1975, **26**, 435.
[36] M. B. Winstead, J. F. Lamb, and H. S. Winchell, *J. Nuclear Medicine*, 1973, **14**, 747.

The initial step in each reaction involved the passage of a stream of $^{11}CO_2$ (to which 0.5—1.0 mmol of carrier was added) into a cold ether solution of 5—10 mmol of the appropriate reagent. Subsequent steps leading to the recovery of material for administration to experimental animals took only some 15—20 min, and so samples of high activity could be obtained. Winstead et al.[36] by these methods have prepared the 11 aliphatic carboxylates, eight benzoic acid derivatives, and seven other carboxylates listed in Table 1.

More recently, Machulla et al.[37] have used a reaction essentially similar to the above in preparing [^{11}C]nicotonic acid ([carboxy-^{11}C]3-pyridine carboxylic acid). High pressure liquid chromatography was used to isolate pure samples of this product, free from [carboxy-^{11}C]valeric acid which was a by-product of the synthesis.

The preparation of [^{11}C]acetate by carboxylation of MeMgBr has been outlined by Lathrop et al.[38] and De Grazia et al.[29] have taken this preparation a step further, obtaining [^{11}C]ethanol ($CH_3{}^{11}CH_2OH$) by reduction of the carboxylated Grignard reagent with $LiAlH_4$. Straatman et al.[39] have used a different approach in synthesizing [1-^{11}C]acetoacetic acid, adding $^{11}CO_2$ to the stable enolate anion of acetone which was generated by the reaction of CH_3Li with isopropenyl acetate.

Synthesis of a ^{11}C-labelled psychoactive drug, chlorpromazine, has been achieved by Comar et al. Initially, chlorpromazine-[^{11}C]methiodide (1) was prepared,[40] by the route shown in Scheme 1.

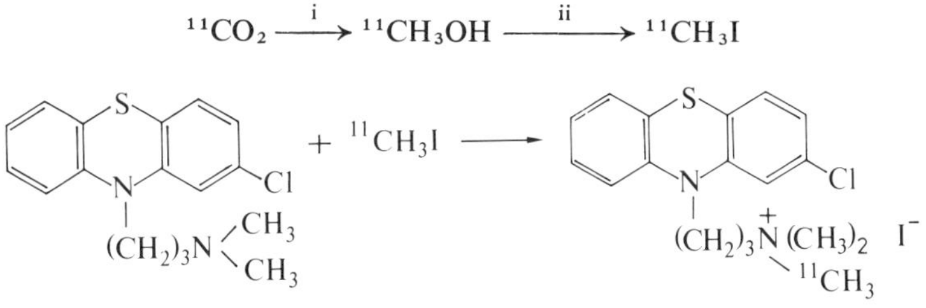

$$^{11}CO_2 \xrightarrow{i} {}^{11}CH_3OH \xrightarrow{ii} {}^{11}CH_3I$$

(1)

Reagents; i, $LiAlH_4$; ii, HI

Scheme 1

Subsequently, the procedure was modified, and [^{11}C]chlorpromazine (2) was synthesized[41,42] (see Scheme 2).

More recently, the use of $^{11}CH_3I$ or $H^{11}CHO$ in labelling several more compounds (e.g. nicotine, valium, caffeine) has been outlined,[43] and the preparation of

[37] H.-J. Machulla, P. Laufer, and G. Stöcklin, *Radiochem. Radioanalyt. Letters*, 1974, **18**, 275.
[38] K. A. Lathrop, P. V. Harper, B. H. Rich, R. Dinwoodie, H. Krizek, N. Lembares, and I. Gloria, ref. 6, p. 471.
[39] M. G. Straatman, A. G. Hortmann, and M. J. Welch, *J. Labelled Compounds*, 1974, **10**, 175.
[40] D. Comar, M. Maziere, and C. Crouzel, ref. 6, Vol. 1, p. 461.
[41] D. Comar, M. Maziere, and C. Raymond in 'Radioaktive Isotop in Klinik und Forschung', ed. R. Höfer, Urban und Schwarzenberg, Munich, 1975, Vol. 11, p. 81.
[42] M. Maziere, J. L. Sainte-Laudy, C. Crouzel, and D. Comar, ref. 7, p. 189.
[43] P. Marche, C. Marazano, M. Maziere, J. L. Morgat, P. de la Llosa, D. Comar, and P. Fromageot, *Radiochem. Radioanalyt. Letters*, 1975, **21**, 53.

Table 1

Structure	Radiochemical yield (%)
Aliphatic carboxylates	
Me$^{11}CO_2$Na	84
Et$^{11}CO_2$Na	98
$CH_2CH^{11}CO_2$Na	54
Pr^n $^{11}CO_2$Na	98
Pr^i $^{11}CO_2$Na	96
Bu^t $^{11}CO_2$Na	68
Bu^n $^{11}CO_2$Na	59

Structure	Radiochemical yield (%)
Me$(CH_2)_4$$^{11}CO_2$Na	65
Me$(CH_2)_5$$^{11}CO_2$Na	40
cyclohexyl-$^{11}CO_2$Na	87
Me$(CH_2)_6$$^{11}CO_2$Na	59

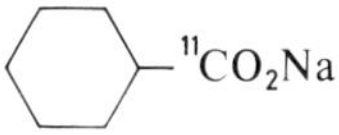

Benzoic Acid Derivatives

Structure	Yield
Ph–$^{11}CO_2$Na	77
4-Cl-C$_6$H$_4$–$^{11}CO_2$Na	82
3,4-(MeO)$_2$-C$_6$H$_3$–$^{11}CO_2$Na	40
4-HO-C$_6$H$_4$–$^{11}CO_2$Na	90
2-HO-C$_6$H$_4$–$^{11}CO_2$Na	50
2-Me-C$_6$H$_4$–$^{11}CO_2$Na	59
3-CF$_3$-C$_6$H$_4$–$^{11}CO_2$Na	22
4-PhO-C$_6$H$_4$–$^{11}CO_2$Na	55

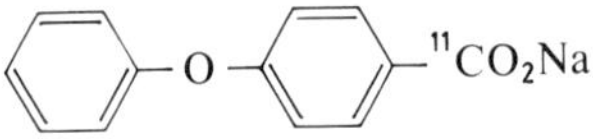

Other carboxylates

Structure	Yield
PhCH$^{11}CO_2$Na	
thiophene-2-$^{11}CO_2$Na	71
camphor-derived $^{11}CO_2$Na (Me,Me,Me, O)	55
1-naphthyl-^{11}CONa	74
acenaphthene-$^{11}CO_2$Na	50
anthracene-9-$^{11}CO_2$Na	95
phenanthrene-$^{11}CO_2$Na	50

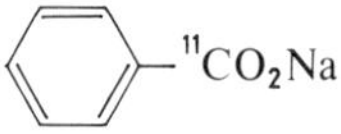

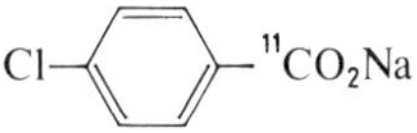

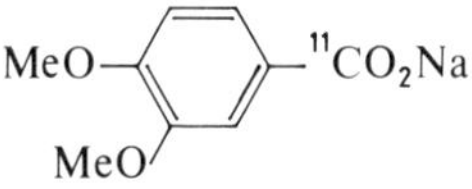

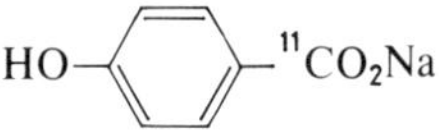

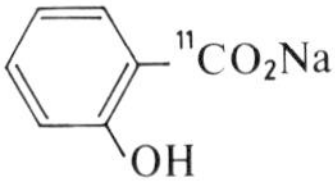

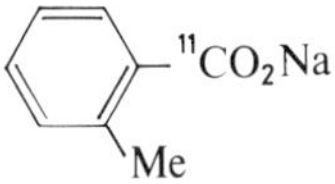

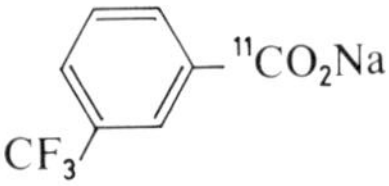

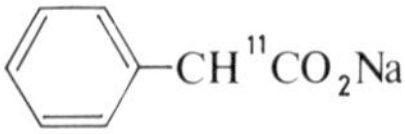

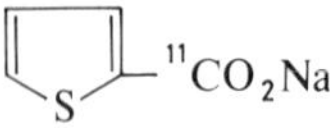

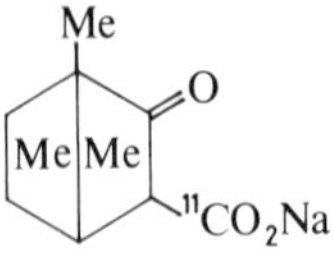

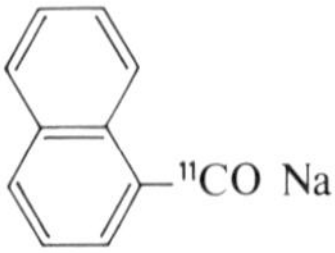

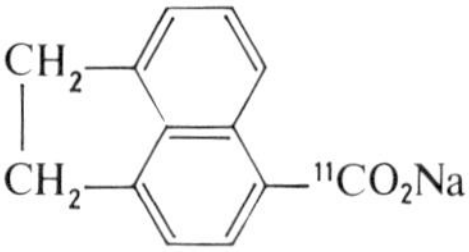

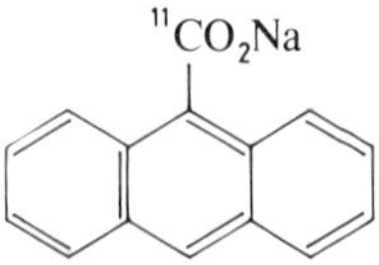

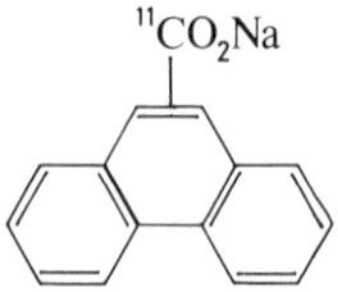

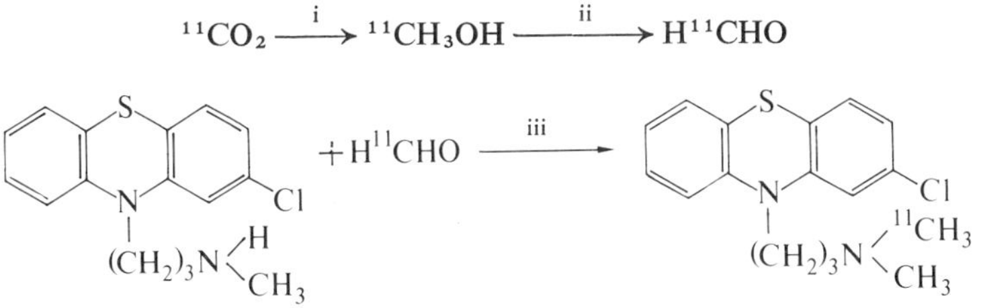

$^{11}CO_2 \xrightarrow{i} {}^{11}CH_3OH \xrightarrow{ii} H^{11}CHO$

[Phenothiazine structure with Cl and (CH₂)₃N(H)(CH₃) side chain] + H^{11}CHO $\xrightarrow{iii}$ [Phenothiazine structure with Cl and (CH₂)₃N(^{11}CH₃)(CH₃) side chain]

(2)

Reagents: i, LiAlH$_4$; ii, Ag, 500 °C; iii, HCO$_2$H, DMF

Scheme 2

^{11}C-methyl-labelled ovine luteinizing hormone has been described in some detail.[44]

The amino-acids valine-^{11}C and alanine-^{11}C are reported to have been prepared by the Strecker synthesis,[45] but a new method of preparing ^{11}C-labelled amino-acids has been developed by Vaalburg.[46,47] The reaction sequence is given in Scheme 3.

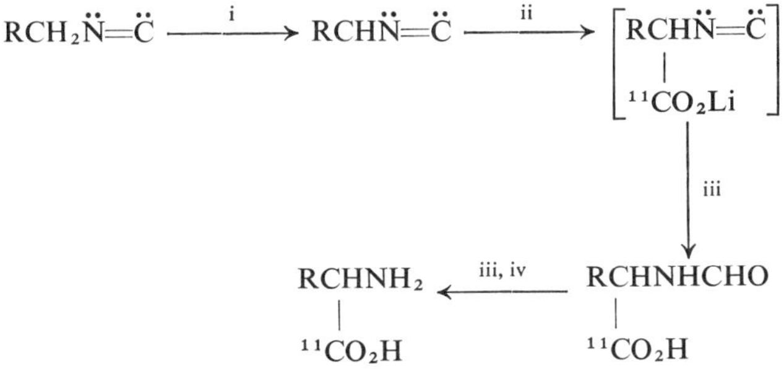

$RCH_2\ddot{N}=\ddot{C} \xrightarrow{i} RCH\ddot{N}=\ddot{C} \xrightarrow{ii} \begin{bmatrix} RCH\ddot{N}=\ddot{C} \\ | \\ {}^{11}CO_2Li \end{bmatrix}$

$\downarrow iii$

$\underset{\underset{{}^{11}CO_2H}{|}}{RCHNH_2} \xleftarrow{iii, iv} \underset{\underset{{}^{11}CO_2H}{|}}{RCHNHCHO}$

Reagents: i, C$_4$H$_9$Li, THF; ii, ^{11}CO$_2$; iii, H$_3$O$^+$; iv, Δ

Scheme 3

Using this method, Vaalburg has synthesized [1-^{11}C]α-phenylglycine and [1-^{11}C]-DL-α-phenylalanine in good yield and within 40 min of introducing ^{11}CO$_2$ in the reaction sequence.

The biosynthesis of [^{11}C]glucose and [^{11}C]fructose by exposure of plant leaves to ^{11}CO$_2$ has been described,[48] and the use of liquid chromatography to separate pure samples of these products from hydrolysed leaf extracts has been reported.[48,49] A similar experimental technique, but using brown or red marine algae in place of green algae or plant leaves led to the preparation of ^{11}C-labelled

---

[44] M. Maziere, C. Marazano, and D. Comar, 'Compte-rendus du XVI Colloque de Medicine Nucleaire', ed. G. Meyneil, G. Gaillard, and D. Isabelle, Bloc-Sante, Clermont-Ferrand, 1974, Vol. 1, p. 58.
[45] A. S. Gelbard, T. Hara, R. S. Tilbury, and J. S. Laughlin, ref. 6, Vol. 1, p. 239.
[46] W. Vaalburg, Thesis, Rijksuniversiteit, Groningen, Netherlands, 1974.
[47] W. Vaalburg, H. D. Beerling-van der Molen, and M. G. Woldring, *Nuclear-Medicine*, 1975, **14**, 60.
[48] R. W. Goulding and A. J. Palmer, *Internat. J. Appl. Radiation Isotopes*, 1973, **24**, 7.
[49] M. Straatman and M. J. Welch, *Internat. J. Appl. Radiation Isotopes*, 1975, **24**, 234.

galactose, glycerol, and mannitol.[50] The liquid chromatography of these, and related compounds has been studied in detail.[51]

Syntheses with $H^{11}CN$.—The syntheses of two ^{11}C-labelled catecholamines (dopamine[52,53] and norepinephrine[52,54] hydrochlorides) free from added carrier, have been described; radiochemical yields are 20—30% (synthesis time 65 min) and 10% (40 min), respectively. The reaction sequences are shown in Scheme 4.

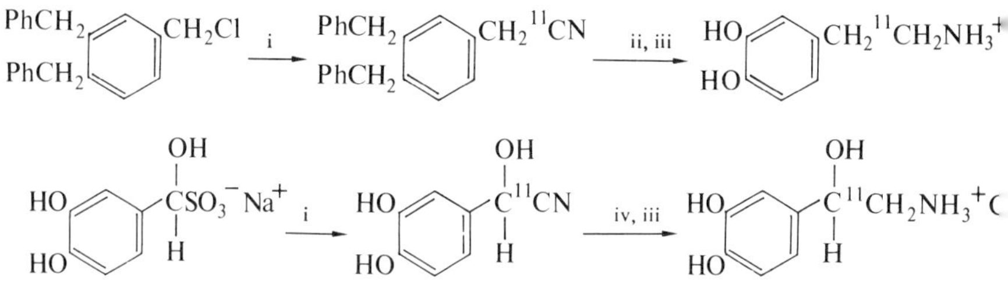

Reagents: i, $Na^{11}CN$ (carrier free); ii, H_2–Pd, iii, HCl; iv, BH_3

Scheme 4

A series of [^{11}C]aminonitriles has been synthesized by Winstead et al.[55] by the reaction of $Na^{11}CN$ with the Schiff bases, Ar—CH=NR(Ar). Ten [^{11}C]-α-N-alkylaminophenylacetonitrile hydrochlorides (Table 2) and twelve [^{11}C]-α-N-arylaminoarylacetonitriles (Table 3) were prepared by the reaction sequences shown.

4 Nitrogen-13

Nitrogen-13, which decays by β^+ emission with a $t_\frac{1}{2}$ of 10 min is the longest-lived radionuclide of nitrogen. Its production, in the form of labelled N_2 gas, by the $^{12}C(d,n)^{13}N$ reaction, using deuterons accelerated in a cyclotron has been studied extensively, and the results have been reported together with a great deal of practical detail.[56] Either solid (graphite or charcoal) or gaseous (CO_2) targets have been employed; the former give higher yields (typically 400 μCi s^{-1}) are more convenient when aqueous solutions of the gas have to be prepared.[56,57] The use of a van de Graaf machine to make ^{13}N–N gas has also been reported.[58]

---

[50] A. J. Palmer and R. W. Goulding, J. Labelled Compounds, 1975, **10**, 627.
[51] R. W. Goulding, J. Chromatog., 1975, **103**, 229.
[52] J. S. Fowler, A. P. Wolf, D. R. Christman, R. R. MacGregor, A. Ansari, and H. Atkins, ref. 7, p. 196.
[53] J. S. Fowler, A. N. Ansari, H. L. Atkins, P. R. Bradley-More, R. R. MacGregor, and A. P. Wolf, J. Nuclear Medicine, 1973, **14**, 867.
[54] J. S. Fowler, R. S. MacGregor, A. N. Ansari, and A. P. Wolf, J. Medicine Chem., 1974, **17**, 246.
[55] M. B. Winstead, P. J. Widner, J. L. Means, M. A. Engstrom, G. E. Graham, A. Khentigan, T. H. Lin, J. F. Lamb, and H. S. Winchell, J. Nuclear Medicine, 1975, **16**, 1049.
[56] J. C. Clark and P. D. Buckingham, ref. 30, Chap. 6, p. 171.
[57] C. Crouzel and D. Comar, Radiochem. Radioanalyt. Letters, 1975, **20**, 273.
[58] R. A. Rydin and R. A. Engler, Phys. Med. Biol., 1974, **19**, 884.

Table 2 *Preparation of ^{11}C-labelled α-N-alkylaminophenylacetonitrile hydrochlorides* (2.5 mmol *scale*)

$$PhCHO + RNH_2 \xrightarrow{i} [PhCH=NR] \xrightarrow{ii, iii} PhCH(^{11}CN)-NHR, HCl$$

Reagents: i, EtOH; ii, aq. Na ^{11}CN, CH_3CO_2H, 0—5 °C, 20 min; iii, HCl gas, ether

R	Radiochemical yield (%)	Chemical yield (%)	Total preparation time/min
Me	39	53	75
$PhCH_2$	18	14	65
Et	65	33	82
$PhCH_2CH_2$	15	6	65
Pr^n	51	39	74
Pr^i	11	26	49
Bu^n	35	40	65
Me_2CHCH_2	33	46	56
$MeCH_2CHMe$	59	—	80
Bu^t	18	27	60

[a] Data from ref. 55.

Table 3 *Preparation of ^{11}C-labelled α-N-arylaminoarylacetonitriles* (5 mmol *reaction scale*)[a]

$$R^1-C_6H_4-CHO + R^2-C_6H_4-NH_2 \xrightarrow{EtOH} R^1-C_6H_4-CH=N-C_6H_4-R^2$$
$$\xrightarrow{\text{aq. Na CN, AcOH, 30—60°C, 10—20 min}} R^1-C_6H_4-CH(^{11}CN)-NH-C_6H_4-R^2$$

R^1	R^2	Radiochemical yield (%)	Chemical yield (%)	Total preparation time/min
H	H	44	41	58
H	Cl	19	20	67
H	Me	61	—	45
H	MeO	48	37	46
H	CO_2H	26	51	60
Me	H	52	—	53
MeO	H	49	—	67
Cl	H	56	—	55
Me	Me	54	66	38
MeO	Me	28	44	35
Cl	Cl	30	36	40
Cl	Me	26	55	75

[a] Data from ref. 55.

The preparation of $^{13}NH_3$ has been investigated by several groups. In earlier work, the $^{12}C(d,n)^{13}N$ reaction was again employed, with deuteron beams of 6—8 MeV incident energy and targets of solid Al_4C_3[59] or gaseous CH_4.[59-61] Both, after chemical processing, yielded $^{13}NH_3$ which was far from pure, low molecular weight amines and cyanide being identified amongst the contaminants.

Better yields of a purer product have been obtained from the $^{16}O(p,\alpha)^{13}N$ reaction, using proton beams of 15—19 MeV incident energy and either gaseous O_2[62,63] or, more successfully, water targets.[62,64-66] Some typical results are shown in Table 4 (taken from ref. 66).

Four ^{13}N-labelled amino-acids have been synthesized by enzyme reactions, using $^{13}NH_3$ as a precursor. The first to be reported[59,65,67] were [^{13}N]-L-glutamine and [^{13}N]-L-glutamic acid, from reactions (3) and (4).

Table 4 *Comparison of $^{13}NH_3$ yields from various production methods*[a]

Nuclear reaction	Target material	Particle energy/MeV	Radiochemical purity (%)	Yield mCi(μA)$^{-1}$ (20 min^{-1})
$^{12}C(d,n)^{13}N$	CH_4	8	80—95	4.4
$^{16}O(p,\alpha)^{13}N$	O_2	<15	99.6	5.6
	H_2O	<15	100	15
	H_2O	19	99.9	36

[a] Data from ref. 66.

$$\text{L-glutamate} + {}^{13}NH_3 \xrightarrow[\text{ATP}]{\text{glutamine synthetase}} [{}^{13}N]\text{-L-glutamine} \quad (3)$$

$$\alpha\text{-ketoglutarate} + {}^{13}NH_3 \xrightarrow[\text{NADH}]{\text{L-glutamic acid dehydrogenase}} [{}^{13}N]\text{-L-glutamic acid} \quad (4)$$

In each preparation, the substrate and cofactors were incubated with $^{13}NH_3$ for up to 15 min, following which the required product was isolated from the mixture by column chromatography within a further 5 min. Up to 90% of the ^{13}N is reportedly incorporated in the product, and samples with activities of 25—60 mCi have been obtained[65] for use in distribution studies in animals.

[59] M. G. Straatman and M. J. Welch, *Radiation Res.*, 1973, **56**, 58.
[60] P. V. Harper, J. Schwartz, R. N. Beck, K. A. Lathrop, N. Lembares, H. Krizek, I. Gloria, R. Dinwoodie, A. McLaughlin, V. J. Stark, C. Bekerman, P. B. Hoffer, A. Gottschalk, L. Resnekov, J. Al-Sadir, A. Mayorga, and H. L. Brooks, *Radiology*, 1973, **108**, 613.
[61] M. E. Read, M. T. McEllistrem, and W. A. Pettit, *Phys. Med. Biol.*, 1975, **20**, 493.
[62] H. Krizek, N. Lembares, R. Dinwoodie, I. Gloria, K. A. Lathrop, and P. V. Harper, *J. Nuclear Medicine*, 1973, **14**, 629.
[63] N. J. Parks, N. F. Peek, and E. Goldstein, *Internat. J. Appl. Radiation Isotopes*, 1975, **26**, 683.
[64] P. V. Harper, K. A. Lathrop, H. Krizek, N. Lembares, and R. Dinwoodie, ref. 7, Chapter 19, p. 180.
[65] A. S. Gelbard, L. P. Clark, J. M. McDonald, W. G. Monahan, R. S. Tilbury, T. Y. T. Kuo, and J. S. Laughlin, *Radiology*, 1975, **116**, 127.
[66] W. Vaalburg, J. A. A. Kamphuis, H. D. Beerling-van der Molen, S. Reiffers, A. Rijskamp, and M. G. Woldring, *Internat. J. Appl. Radiation Isotopes*, 1975, **26**, 316.
[67] M. B. Cohen, L. Spolter, N. S. MacDonald, D. T. Masuoka, S. Laws, H. H. Neely, and J. Takahashi, ref. 6, Vol. 1, p. 483.

A similar experimental approach has been employed in making [^{13}N]-L-asparagine,[68] the reaction being:

$$\text{L-aspartic acid} + {}^{13}\text{NH}_3 \xrightarrow[\text{ATP}]{\text{asparagine synthetase}} [{}^{13}\text{N}]\text{-L-asparagine} \qquad (5)$$

The preparation of [^{13}N]-L-alanine has also been described,[69] according to the reaction:

$$[{}^{13}\text{N}]\text{-L-glutamic acid} \xrightarrow{\text{glutamic-pyruvic transaminase}} [{}^{13}\text{N}]\text{-L-alanine} \qquad (6)$$

Again, the product was isolated by column chromatography. An improved method of synthesizing [^{13}N]-L-glutamic acid and [^{13}N]-L-alanine, by the above reactions, makes use of enzymes immobilized on porous derivatized silica beads.[70]

5 Oxygen-15

Oxygen-15, which decays by β^+ emission, is the longest-lived radionuclide of oxygen, with a half-life of only 2.1 min. Not surprisingly, although several applications have been found for the radionuclide in simple molecules (O_2, CO, CO_2, H_2O) no complex syntheses have been reported.

Only one nuclear reaction ^{14}N(d,n)^{15}O is reported[71,72] to have yielded useful quantities of this radionuclide. Deuterons of 5—6 meV incident energy have been used, with nitrogen gas as the target material.

Systems enabling the ^{15}O to be recovered continuously in different chemical forms have been described[71] with extensive practical details. If a small quantity ($<4\%$) of O_2 is added to the N_2 flowing through the target vessel, $>95\%$ of the ^{15}O appears in the gas stream as ^{15}OO. If, instead of O_2, about 2% CO_2 is added, $>99\%$ of the ^{15}O appears as C^{15}OO.

C^{15}O is conveniently prepared by 'on-line' reduction of ^{15}OO by activated charcoal at 900 °C, whereas $H_2{}^{15}$O is prepared by addition of excess H_2 to the gas stream prior to passage through a heated palladium catalyst.

6 Fluorine-18

For many years ^{18}F, administered as an aqueous carrier-free fluoride solution, aided the diagnosis of bone disease because it accumulated to a higher level in diseased bone than in normal tissues. More recently, however, its physical properties (110 min half-life, and readily detectable β^+ emission) coupled with the chemical properties of fluorine (e.g. the high dissociation energy, 107 kcal mol^{-1}, of the C—F bond) have led to the recognition of ^{18}F as a useful label for a wide range of organic compounds. The arguments for and against the use of ^{18}F for this purpose have been

---

[68] A. S. Gelbard, L. P. Clarke, and J. S. Laughlin, *J. Nuclear Medicine*, 1974, **15**, 1223.
[69] M. B. Cohen, L. Spolter, N. S. MacDonald, C. C. Chang, and J. Takahashi, ref. 7, Chapter 20, p. 184.
[70] M. B. Cohen, L. Spolter, C. C. Chang, N. S. MacDonald, J. Takahashi, and D. D. Babinet, *J. Nuclear Medicine*, 1974, **15**, 1192.
[71] J. C. Clark and P. D. Buckingham, ref. 30, Chapter 5, p. 122.
[72] L. W. Cress and R. A. Rydin, *Phys. Med. Biol.*, 1973, **18**, 742; *ibid.*, 1974, **19**, 884.

discussed by Wolf et al.[5] and by Robinson,[73] in reviews of earlier work on the preparation of ^{18}F-labelled compounds.

Several nuclear reactions have been used to prepare ^{18}F. Effective routine production in a nuclear reactor has been described[74] which utilizes the ^{6}Li(n,α)^{3}H, ^{16}O(^{3}H,n)^{18}F reactions. About 2 g Li$_2$CO$_3$ (^{6}Li enriched to >95%) is irradiated in a quartz container for 3—4 h in a flux of 1.10^{13} n cm^{-2} s^{-1} and the ^{18}F (average yield about 14 mCi) is recovered by distillation in Teflon apparatus. Better yields (65—75 mCi) have been obtained[75] by replacing the quartz container by a graphite-lined aluminium can. A third method[76] achieves much more rapid recovery of the ^{18}F by irradiation of an aqueous slurry, or paste, of Li$_2$CO$_3$.

Earlier methods of production using accelerators have been reviewed[14] and these, giving generally much higher yields than reactors, are especially advantageous when syntheses with ^{18}F are to be attempted.

Aqueous ^{18}F solutions have long been prepared, using the ^{16}O(α,pn)^{18}F reaction, in 'static' water targets. In a variation on this theme, Lindner et al.[22,77] irradiated water circulating through a stainless steel target vessel exposed to 52 MeV α-particles in a cyclotron. Downstream from the target, an anion-exchange column stripped the ^{18}F produced from the solution. At the end of the irradiation ^{18}F could be recovered from the column by elution with an isotonic saline solution.

The ^{20}Ne(d,α)^{18}F reaction is also recognized as a prolific source of ^{18}F and a 5.4 MeV van de Graaf accelerator, operating at a beam current of only 2 μA for 1 h, has produced up to 30 mCi ^{18}F by this route.[61,78]

Several other reactions yielding ^{18}F from gaseous O$_2$ and Ne targets have been studied in detail (Table 5) and excitation function curves have been published.[79] One of the advantages accruing from the use of dry gases as target materials is that

Table 5 *Reactions used for the preparation of fluorine-18*[a]

Bombardment	Predominant reaction	Q-value/MeV	Threshold energy/ MeV
O + t	^{16}O(t,n)^{18}F	+1.270	0
O + ^{3}He	^{16}O(^{3}He,p)^{18}F	+2.003	0
	^{16}O(^{3}He,n)^{18}Ne→^{18}F	−3.196	3.795
O + α	^{16}O(α,pn)^{18}F	−18.544	23.180
	^{16}O(α,2n)^{18}Ne→^{18}F	−23.773	29.716
Ne + d	^{20}Ne(d,α)^{18}F	+2.796	0
Ne + ^{3}He	^{20}Ne(^{3}He,αp)^{18}F	−2.697	3.102
	^{20}Ne(^{3}He,αn)^{18}Ne→^{18}F	−7.296	9.115

[a] Data from ref. 79.

[73] G. D. Robinson, ref. 7, Chapter 14, p. 141.
[74] F. Helus, O. Krauss, and W. Maier-Borst, *Radiochem. Radioanalyt. Letters*, 1973, **15**, 225.
[75] P. K. H. Chan, G. Firnau, and E. S. Garnett, *Radiochem. Radioanalyt. Letters*, 1974, **19**, 237.
[76] W. C. Parker, C. P. G. da Silva, and W. H. G. Francis, Proceedings 7th International Hot Atom Chemistry Symposium, Jülich, West Germany, 1974.
[77] L. Lindner, T. H. G. A. Suer, G. A. Brinkman, and J. Th. Beenboer, *Internat. J. Applied. Radiation Isotopes*, 1973, **24**, 124.
[78] M. F. Reed, M. T. McEllistram, D. F. Preston, and R. H. Beihm, *Radiation Research*, 1973, **55**, 570.
[79] T. Nozaki, M. Iwamoto, and T. Ido, *Internat. J. Appl. Radiation Isotopes*, 1974, **25**, 393.

Preparation of Radiopharmaceuticals and Labelled Compounds 87

the ^{18}F may be recovered in an anhydrous form, and this is of great value in many syntheses.

Studies have been made of the direct interaction of ^{18}F atoms, from the ^{19}F(n,2n) reaction, with a variety of inorganic and organic compounds.[80—82] Such experiments have all been carried out at low (*ca.* 1 µCi) radioactivity levels, however, and the authors acknowledge[80] that at the higher levels necessary to achieve syntheses of useful quantities of labelled compounds, radiation chemical effects could significantly alter the nature of the products.

^{18}F-Intermediates.—Although aqueous ^{18}F, prepared from either reactor or cyclotron irradiations, has been used directly with some success in the synthesis of certain amino-acids and other compounds,[83—87] many fluorination reactions will proceed only in anhydrous conditions. The preparation of anhydrous ^{18}F intermediate compounds for use in such reactions has therefore received the attention of several groups of workers.

Lambrecht and Wolf[88,89] have prepared anhydrous [^{18}F]-F$_2$ (in batches of > 0.5 Ci) by deuteron bombardment of neon gas under pressure in a fluorine-treated nickel target vessel. Clark *et al.*[90] prepared Ag^{18}F and Ag^{18}F$_2$ by the same reaction in a target vessel lined with AgF of AgF$_2$. The same group[91] labelled several more non-volatile intermediates, containing ionic or potentially ionic fluorine atoms (*e.g.* KF, SbF$_3$, and several diazonium fluoroborate compounds) by circulating neon through a glass-lined target vessel and trapping the ^{18}F formed (probably as NO^{18}F or ·O$_2$^{18}F, and not as ^{18}F·) on a filter, downstream from the target, on which 5—10 mg of the appropriate compound were supported.

Robinson[92] has prepared dry [^{18}F]fluoride-labelled anion exchange resin by circulating an aqueous ^{18}F solution through a small ampoule containing the resin, then rinsing it with methanol and ether before drying under vacuum at 40 °C. Mantescu *et al.*[93] have made anhydrous K^{18}F, CH$_3$CO$_2$H solutions by dissolving reactor-irradiated Li$_2$CO$_3$ in glacial acetic acid, absorbing ^{18}F from this solution on a cellulose column, and eluting it with acetone containing KF. After evaporation of the acetone, the K^{18}F was redissolved in more glacial acetic acid.

Dry K^{18}F has also been prepared by de Kleijn *et al.*[94] from an aqueous solution of reactor-irradiated Li$_2$CO$_3$ which was treated with a strongly acidic cation-

---

[80] F. S. Rowland, J. A. Cramer, R. S. Iyer, R. Millstein, and R. L. Williams, ref. 6, Vol. 1, p. 383.
[81] J. A. Cramer and F. S. Rowland, *J. Amer. Chem. Soc.*, 1974, **96**, 6579.
[82] R. S. Iyer and F. S. Rowland, *Chem. Phys. Letters*, 1973, **21**, 346.
[83] A. J. Palmer, J. C. Clark, R. W. Goulding, and M. Roman, ref. 6, Vol. 1, p. 291.
[84] S. Garnett and G. Firnau, ref. 6, Vol. 1, p. 405.
[85] G. Firnau, C. Nahmias, and S. Garnett, *Internat. J. Applied Radiation Isotopes*, 1973, **24**, 182.
[86] G. Firnau, C. Nahmias, and S. Garnett, *J. Medicin. Chem.*, 1973, **16**, 416.
[87] C. S. Kook, M. F. Reed, and G. A. Digenis, *J. Medicin. Chem.*, 1975, **18**, 533.
[88] R. M. Lambrecht and A. P. Wolf, ref. 6, Vol. 1, p. 275.
[89] R. M. Lambrecht and A. P. Wolf, ref. 7, Chapter 11, p. 111.
[90] J. C. Clark, R. W. Goulding, and A. J. Palmer, ref. 6, Vol. 1, p. 411.
[91] J. C. Clark, R. W. Goulding, M. Roman, and A. J. Palmer, *Radiochem. Radioanalyt. Letters*, 1973, **14**, 101.
[92] G. D. Robinson, ref. 6, Vol. 1, p. 423.
[93] C. Mantescu, A. Genunche, and L. Simonescu, ref. 6, Vol. 1, p. 395.
[94] J. P. de Kleijn, H. J. Meeuwissen, and B. van Zanten, *Radiochem. Radioanalyt. Letters*, 1975, **23**, 139.

exchange resin, to decompose the carbonate anions and absorb the lithium cations. The required amount (0.1—2.4 mmol) of KF was added after this process, and the solution evaporated to dryness. Within 80 min of the end of irradiation, >90% of the original ^{18}F activity was present in dry K^{18}F.

Preparation of Specific ^{18}F-Organic Compounds.—Probably the reaction most frequently used so far to introduce a ^{18}F atom into an organic compound is the Schiemann degradation, thermal decomposition of a diazonium tetrafluoroborate compound:

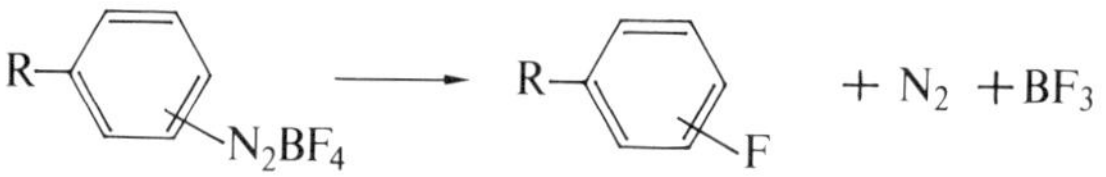

$$R-\text{C}_6\text{H}_4-N_2BF_4 \longrightarrow R-\text{C}_6\text{H}_4-F + N_2 + BF_3 \quad (7)$$

An advantage of this method is the relative ease with which ^{18}F atoms can be incorporated in the precursor by exchange reactions. However, a major disadvantage is the loss of 75% of the activity so incorporated as B^{18}F$_3$.

Several labelled amino-acids have been prepared by this general method, including DL-mixtures of [^{18}F]-m- or -p-fluorophenylalanine,[91] [^{18}F]-3-fluorotyrosine,[83,91] [^{18}F]-5- and -6-fluorotryptophan,[91] and [^{18}F]-5-fluorodopa.[84-86] Recently, the synthesis of the pure L-isomer of [^{18}F]-p-fluorophenylalanine has been described.[95] The reaction sequence (Scheme 5) is typical of the earlier [^{18}F]fluoroamino-acid syntheses, to the point where the mixture of the D- and L-acylated fluoroamino-acid, (3), is reached. Hydrolysis of the L-, but not the D- form of (3) was effected by the enzyme L-amino acylase (Scheme 5).

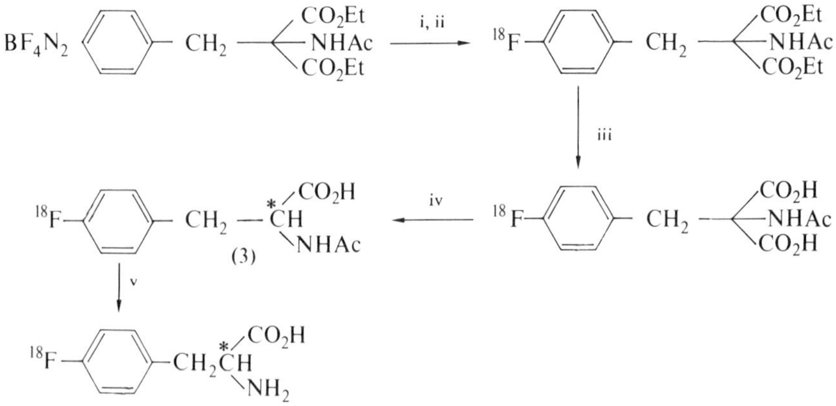

(4) [^{18}F]-L-p-fluorophenylalamine

Reagents: i, Heterogeneous exchange with ^{18}F (ref. 91); ii, Δ, 180 °C; iii, OH$^-$; iv, H$_3$O$^+$; v, L-aminoacylase

Scheme 5

[95] R. W. Goulding and S. W. Gunasekera, *Internat. J. Appl. Radiation Isotopes*, 1975, **26**, 561.

Isolation of the required product, (4), from the reaction mixture was achieved by liquid chromatography.

Schiemann degradation of the diazonium tetrafluoroborate of 4-[4-(p-chlorophenyl)-4-hydroxypiperidino]-4′-aminobutyrophenone, following exchange with ^{18}F in aqueous acetone, has been used in the preparation[87] of [^{18}F]halogenoperidol (5).

$$^{18}F-\underset{}{\text{C}_6\text{H}_4}-\overset{O}{\underset{||}{\text{C}}}\text{CH}_2\text{CH}_2\text{CH}_2\text{N}\underset{}{\bigg\langle}\underset{\text{C}_6\text{H}_4\text{Cl}}{\overset{\text{OH}}{\bigg|}}$$

(5)

The preparation of some [^{18}F]fluorocarboxylates ([^{18}F]-fluoroacetate, -fluorohexanoate, and -fluorotetradecanoate) and derivatives has been reported.[92] The general procedure is to add a bromocarboxylate ester (ca. 10 mg) to dry ^{18}F-labelled anion exchange resin in a small ampoule which is then sealed and heated to 180 °C for up to 90 min. Interhalogen exchange occurs, and the products are subsequently separated by preparative gas–liquid chromatography. Hydrolysis of the ^{18}F-labelled esters yields the required ^{18}F-carboxylate. [2-^{18}F]Fluoroethanol has also been prepared, by a modification of the foregoing method. [2-^{18}F]Ethyl-2-fluoropropionate was synthesized[94] by reaction of the corresponding bromocompound with dry $K^{18}F$ in acetamide, and isolated by gas–liquid chromatography, whilst treatment of tosyl chloride with $K^{18}F$ in the presence of water yielded toluene-p-sulphonyl [^{18}F]fluoride which was isolated by sublimation.[94]

Two fluorocarbons used as aerosol propellants, trichlorofluoromethane ('Freon-11') and dichlorodifluoromethane ('Freon-12') were synthesized, incorporating ^{18}F, by Clark et al.[90] for use in pharmacodynamic studies. The fluorinating agent was $Ag^{18}F$ or $Ag^{18}F_2$, prepared as a coating on the inside walls of a cyclotron target vessel in which the $^{20}Ne(d,\alpha)^{18}F$ reaction had been induced. After irradiation, the neon was pumped out and the appropriate organic reactant (ca. 2g) introduced. The vessel was then sealed and heated for 40—50 min, when a ^{18}F-labelled fluorocarbon was produced by one of the following heterogeneous gas-phase reactions:

$$2\ Ag^{18}F + CCl_3Br \xrightarrow{70\,°C} AgBr, Ag^{18}F + CCl_3{}^{18}F$$

$$2\ Ag^{18}F + CCl_4 \xrightarrow{160—170\,°C} AgCl, Ag^{18}F + CCl_3{}^{18}F$$

$$2\ Ag^{18}F_2 + 2CCl_3F \xrightarrow{160—170\,°C} 2\ Ag^{18}F + 2\ CCl_2{}^{18}FF_2 + Cl_2$$

Chemical yields of up to 90% and radiochemical yields of up to 40% were obtained. Preparative gas–liquid chromatography was used to isolate the required product.

The synthesis of [^{18}F]-5-fluorouracil in high specific activity (ca. 1 mCi mg^{-1}) has been achieved[96] by the direct reaction of anhydrous [^{18}F]-F_2 with uracil in

---

[96] J. S. Fowler, R. D. Finn, R. M. Lambrecht, and A. P. Wolf, *J. Nuclear Medicine*, 1973, **14**, 63.

trifluoracetic acid. This product was purified by sublimation followed by ion-exchange chromatography. Both synthesis and purification were completed within 35 min from the end of the cyclotron irradiation.

7 Bromine-77

Although ^{80m}Br ($t_{\frac{1}{2}} = 4.4$ h) and ^{82}Br ($t_{\frac{1}{2}} = 35.3$ h) both fall within our definition of short-lived radionuclides, neither has found much favour as a label for compounds or radiopharmaceuticals, probably because of the high abundance of relatively high energy γ-rays associated with their decay. (The nuclear recoil chemistry of both these radionuclides has been studied in some depth, but the reader is referred elsewhere[97] for reviews of this subject.)

On the other hand, ^{77}Br, besides being the longest-lived bromine radionuclide ($t_{\frac{1}{2}} = 57.04$ h),[98] decays almost entirely by electron capture (99%) followed by emission of 242 keV (30%) or 520 keV (24%) γ-rays, which make it a quite attractive radionuclide for labelling purposes.[99] Its production by means of the ^{75}As(α,2n) reaction has been studied in considerable detail. The cross-section for this reaction peaks at 26 MeV.[100] Experimental thick target yields have been measured[101] for targets of As metal, As_2O_3, and As_2O_5 [171, 290, and 160 μCi (μAh)$^{-1}$, respectively] and compared to the theoretical yields [498, 293, and 216 μCi (μAh)$^{-1}$] for 1 h bombardments with beam currents limited by target stability. From these results, and other considerations, As_2O_3 was the recommended target material for routine production of ^{77}Br, and a method of recovering the radionuclide as carrier-free bromide was described.

Higher yields of ^{77}Br are claimed[102] to be obtained from Se(p,xn) or (d,xn) reactions, and its recovery from selenium targets has been discussed. Indirect production, through the sequence

$$^{79}\text{Br}\,(\alpha,6n)\,^{77}\text{Rb} \xrightarrow{\text{short}} {}^{77}\text{Kr} \xrightarrow{1.2\,\text{h}} {}^{77}\text{Br}$$

has also been investigated[103] and may be advantageous if it proves possible to prepare ^{77}Br-labelled compounds by exposure to the ^{77}Kr intermediate ('excitation labelling').[104] However, a major disadvantage of this method is the low yield of ^{77}Kr obtained with α-particle beams of less than about 95 MeV.

The potential advantages of ^{77}Br over iodine radionuclides as a protein-label (which include greater *in vitro* stability) have been discussed, and the preparation of three model compounds, [^{77}Br]-tyrosine, -albumin and -thyroglobulin, has been described.[99] Under optimum conditions, about 80% of the ^{77}Br used was incor-

---

[97] D. S. Urch, in 'Radiochemistry' Vol. 2, ed. G. W. A. Newton (Specialist Periodical Reports), The Chemical Society, London, 1975, p. 1; P. Glentworth and A. Nath, *ibid.*, p. 74.
[98] S. L. Waters and M. J. Woods, *Internat. J. Appl. Radiation Isotopes*, 1975, **26**, 484.
[99] L. Knight, K. A. Krohn, M. J. Welch, B. Spomer, and L. P. Hagar, ref. 7, Chapter 15, p. 149.
[100] S. L. Waters, A. D. Nunn, and M. L. Thakur, *J. Inorg. Nuclear Chem.*, 1973, **35**, 3413.
[101] A. D. Nunn and S. L. Waters, *Internat. J. Appl. Radiation Isotopes*, 1975, **26**, 731.
[102] B. Z. Iofa and Yu. G. Sevast'yanov, *Soviet Radiochem.*, 1975, **16**, 555.
[103] F. Helus, W. Maier-Borst, R. M. Lambrecht, and A. P. Wolf, in 'Proc. 7th International Conference on Cyclotrons and their Applications', ed. W. Joho, Birkhauser, Basel, 1975, p. 474.
[104] R. M. Lambrecht and A. P. Wolf, ref. 7, Chapter 11, p. 111.

porated in the labelled product, from which the free ^{77}Br bromide was removed by liquid chromatography.

The psychodysleptic drug 2,5-dimethoxyphenylisopropylamine (DPIA) is reported[105,106] to have been labelled with ^{77}Br or ^{82}Br, to produce [4-Br]DPIA (6) by direct bromination of DPIA with ^{77}Br$_2$ or ^{82}Br$_2$ in acid solution.

$$Br\text{-}\underset{MeO}{\overset{OMe}{C_6H_2}}\text{-}CH_2\text{-}\underset{NH_2}{\overset{Me}{CH}}$$

(6)

8 Iodine-123

The chemistry of iodine is such that a wide variety of compounds can be labelled with radioactive iodine atoms relatively easily, either by direct iodination or by exchange reactions. For many years ^{131}I($t_{\frac{1}{2}}$ = 8 d), which is cheap and readily available, was extensively and almost exclusively used for this purpose. More recently, in applications such as radioimmunoassay, where a longer-lived label is preferable, ^{125}I($t_{\frac{1}{2}}$ = 60 d) has come into prominence. However, if the labelled substance is to be used *in vivo*, then with few exceptions the short-lived ^{123}I ($t_{\frac{1}{2}}$ = 13.2 h) is the radionuclide of choice.

Iodine-123 can be produced by charged-particle irradiation of various target materials, either directly, when it is the primary product of a nuclear reaction, or indirectly, when it is obtained from the decay of its precursor, ^{123}Xe($t_{\frac{1}{2}}$ = 2.1 h). Methods of production which had been investigated up to the end of 1973 have been reviewed by Weinreich *et al.*,[107] and by Sodd,[108] who points out two advantages of the indirect procedures. First, because of competing nuclear reactions it is impossible to produce ^{123}I directly free from measurable quantities of undesirable radioiodine contaminants, notably ^{124}I($t_{\frac{1}{2}}$ = 4.2 d). Although this problem can be reduced by the use of highly isotopically enriched target materials, it cannot be entirely eliminated by this means. However, the problem is effectively overcome when ^{123}I is produced from its ^{123}Xe precursor, since the latter may be so easily separated from directly formed iodine radionuclides. Second, the decay of ^{123}Xe yields monatomic ^{123}I species which have sufficient chemical reactivity to cause labelling reactions with organic molecules.

Direct Production of ^{123}I.—Of the direct methods of production, only the ^{124}Te(p, 2n)^{123}I route seems to have attracted much recent attention, probably because highly enriched ^{124}Te is available at a more reasonable cost than other tellurium isotopes. Thus, using a 25 MeV proton beam from the isochronous cyclotron of Louvain University, and a target of 300 μg cm^{-2} ^{124}Te, enriched to 93.4%,

---

[105] T. Sargent, D. A. Kalbhen, A. T. Shulgin, G. Braun, H. Stauffer, and N. Kusubov, *Neuropharmacology*, 1975, **14**, 165.
[106] T. Sargent, D. A. Kalbhen, A. J. Shulgin, H. Stauffer, and N. Kusubov, *J. Nuclear Medicine*, 1975, **16**, 243.
[107] R. Weinreich, O. Schult and G. Stöcklin, *Internat. J. Appl. Radiation Isotopes*, 1974, **25**, 535.
[108] V. J. Sodd, ref. 7, Chapter 12, p. 125.

Cauwe et al.[109] obtained 38 mCi(μAh)$^{-1}$ from a 1 h bombardment. Based on this experimental figure, they now predict a thick target yield of 40 mCi (μAh)$^{-1}$ (30 MeV protons) which compares with an earlier[110] calculated yield of 25 mCi (μAh)$^{-1}$ for this method.

The same reaction has been investigated by Acerbi et al. in Milan.[111] Excitation functions for (p,xn) reactions using natural and enriched tellurium targets were measured. Based on the results obtained, the conditions selected for full scale production purposes call for the use of a powdered Te metal target (91.9% ^{124}Te) 11 MeV, or 930 mg cm^{-2}, thick, and a proton beam of 28 MeV incident energy. A 3 h bombardment of such a target with a 3 μA beam yields 400 mCi of ^{123}I with about 1% ^{124}I at end of bombardment; this is equivalent to about 45 mCi (μAh),$^{-1}$ and in good agreement with the previous authors'[109] expectations.

Acerbi et al.[111] describe in considerable detail the construction of their target system, irradiation technique, and the 'wet chemistry' required to recover the ^{123}I produced as well as the valuable enriched ^{124}Te target material. They found it was necessary to add 10 mg KI as carrier in order to recover 85% of the radioiodine. Despite careful precautions, only 60—70% of the ^{124}Te was recovered from each preparation.

A more recent paper[112] by the same authors describes a 'dry distillation' technique which improves the recovery of the ^{123}I produced, and leaves the ^{124}Te target material intact. After irradiation, the target is placed in a glass tube inside an oven where it is heated in a stream of N$_2$ or He to 300—430 °C (just below the m.p. of Te, 452 °C). The carrier-free radioiodine diffuses out of the heated Te and is swept along in the gas stream (20—40 cm^3 min^{-1}) through a cooling loop, to a trap containing 0.001 mol l^{-1} NaOH where Na^{123}I is presumed to be formed. The authors point out that further development of their approach should enable the ^{123}I to be recovered from a Te target continuously during the actual cyclotron irradiation, and include an outline of the system in which they propose to test this possibility.

Indirect Production of ^{123}**I.**—The principle of 'on-line' recovery has been applied already in indirect methods of ^{123}I production, where, as previously mentioned ^{123}Xe is the primary product required. Constructional, and operational, details of a suitable target system in which to bring about the ^{122}Te(^{4}He,3n)^{123}Xe reaction have been described.[113] This consists of a water-cooled vessel which contains 1—1.5 g of Te metal powder (^{122}Te enriched to 96%) in a compartment through which He gas flows during irradiation with 42 MeV α-particles at currents of about 50 μA. The volatile reaction products (^{123}Xe, together with some ^{125}Xe and directly formed iodine radionuclides from competing nuclear reactions) are swept out in the He gas which, down-stream from the target vessel, passes through a trap cooled to

---

[109] F. Cauwe, J. P. Deutsch, D. Favert, R. Prieels, and M. Cogneau, *Internat. J. Applied Radiation Isotopes*, 1974, **25**, 187.
[110] Ch. Deglume, J. P. Deutsch, D. Favert, and R. Prieels, *Internat. J. Appl. Radiation Isotopes*, 1973, **24**, 291.
[111] E. Acerbi, C. Birattari, M. Castiglioni, F. Resmini, and M. Villa, *Internat. J. Appl. Radiation Isotopes*, 1975, **26**, 741.
[112] E. Acerbi, C. Birattari, M. Castiglioni, and F. Resmini, ref. 103, p. 461.
[113] V. J. Sodd, J. W. Blue, K. L. Sholz, and M. C. Oselka, *Internat. J. Appl. Radiation Isotopes*, 1973, **24**, 171.

$-79\,°C$, which removes the directly produced radioiodine, then through a further trap cooled to $-196\,°C$ which retains the ^{123}Xe and ^{125}Xe. At the end of the irradiation the latter trap is sealed, allowed to reach normal temperature, and set aside, conveniently for about 5 h, to allow for in-growth of ^{123}I. After this, the trap is flushed with gas to remove residual ^{123}Xe and ^{125}Xe, and then rinsed with dilute NaOH to recover the ^{123}I. By this means, ^{123}I yields in the range of 2—4 mCi h^{-1} of irradiation have been obtained, with contamination of about 0.4% ^{125}I, and a trace (0.001%) of directly formed ^{124}I.

A comparison has been made[114] of the yields of ^{123}Xe (and thence ^{123}I) obtained from three enriched tellurium targets (^{122}Te, ^{123}Te, and ^{124}Te) irradiated with ^{3}He or ^{4}He beams of different energies. All the methods were shown to give good yields of high purity ^{123}I, but if the cost of the target material is also taken into account, the $^{124}Te(^{3}He,4n)^{123}Xe$ reaction is considered to be attractive, especially where ^{3}He beams of more than 40 MeV are available.

The disadvantage of having to use expensive enriched tellurium targets can be avoided if proton or deuteron beams, of at least about 50 MeV or 70 MeV respectively, are available, for then monoisotopic iodine can be used as target material. Wilkins et al.[115] have published the excitation functions for the $^{127}I(p,5n)^{123}Xe$ and $^{127}I(p,3n)^{125}Xe$ reactions for protons of 45 to 65 MeV, and find that the former has a peak value of about 500 mbarn at about 55 MeV, whilst the latter falls from about 300 to 150 mbarn. Using these results, optimum yields of ^{123}I have been calculated for several possible target materials (I_2, LiI, NaI, KI, and CH_2I_2) of different thicknesses.

Weinreich et al.[107] have measured the excitation functions of the relevant $^{127}I(d,xn)$ and $^{127}I(d,pxn)$ reactions in order to ascertain the optimum conditions for ^{123}I production via the $^{127}I(d,6n)^{123}Xe$ reaction. The latter has a peak value of about 300 mbarn at 70 MeV, and the authors point out that deuteron beams of this energy are somewhat more frequently available than 50 MeV protons. A sodium iodide gas-flow target arrangement for the routine production of ^{123}I by this route is described in some detail, and practical ^{123}I yields of 8 mCi $(\mu Ah)^{-1}$, with an ^{125}I-impurity level of < 0.2%, are claimed.

^{123}**I-Labelled Compounds.**—The reader is again referred to articles by Wolf et al.[5] and Sodd[108] for reviews of the earlier work on this subject. Even now, ^{123}I is only available on a very limited scale, so it is perhaps not surprising that only a handful of ^{123}I-labelled compounds have so far been described, and that most are direct analogues of well-known ^{131}I-labelled compounds.

The starting point for most of these compounds is Na^{123}I, since it is in this form that ^{123}I is usually recovered, at any rate following direct methods of production. Labelling is then achieved either by halogen exchange, or by direct iodination reactions, following oxidation of $^{123}I^-$ by an appropriate method.

One of the most widely utilized ^{123}I-compounds is the sodium salt of [^{123}I]-o-iodohippuric acid ('[^{123}I]hippuran') (7).

---

[114] M. Guillaume, R. M. Lambrecht, and A. P. Wolf, *Internat. J. Appl. Radiation Isotopes*, 1975, **26**, 703.
[115] S. R. Wilkins, S. T. Shimose, H. H. Hines, J. A. Jungerman, F. Hegedus, and G. L. De Nardo, *Internat. J. Appl. Radiation Isotopes*, 1975, **26**, 279.

$$\text{o-}^{123}\text{I-C}_6\text{H}_4\text{-CONHCH}_2\text{CO}_2\text{H}$$

(7)

An early method of preparation has been described[116] in which halogen exchange occurs in aqueous solution. Appropriate quantities of $^{123}\text{I}^-$, I_2, diethanolamine, and hippuran were heated under reflux for 2 h to achieve 70% incorporation of ^{123}I in the product, which was recovered by precipitation following acidification by conc. HCl and cooling in a salt–ice mixture. Residual free iodide and other reactants were removed by washing the precipitate with ice-cold water before the labelled product was redissolved in dilute NaOH.

A more efficient, and much faster method of preparing [^{123}I]hippuran has been described[117] more recently. Depending upon the specific activity required, 5—200 mg hippuran, slurried in 1.5 cm^3 ethanol, was added to a test tube containing 1—5 cm^3 0.1 cm^3 NaOH, 5—10 mCi $^{123}\text{I}^-$, and 10—20 μg I$^-$ carrier. The tube was placed in an oil bath, previously heated to 180 °C and the solution evaporated to dryness under a stream of N_2. The residue was allowed to melt (m.p. 170—175 °C) and about 1 min later the tube was withdrawn from the oil bath. After cooling, the residue was redissolved in dil. NaOH, from which the product was precipitated by conc. HCl and purified by the method previously reported. More than 90% of the ^{123}I was found in the product; this was established by an ascending paper-chromatographic method.

Robinson and Lee[118] have described the preparation of some [^{123}I]fatty acids in which the ^{123}I atom occupies a volume and position in the molecule similar to a terminal methyl-group. Oleic, linoleic, and linolinic acids were iodinated in ether solution by addition of ^{123}ICl across a single unsaturated bond. The special miniature apparatus used for the rapid preparation of high specific activity products is described in some detail. 16-Iodo-9-hexadecoic, 11-iodo-undecanoic, and 6-iodohexanoic acids were also prepared, again in ether solution, but by interhalogen replacement of Br in the corresponding bromocarboxylic acids. After reaction for 3—4 h the solvent was evaporated and the hydrophobic acids were recovered by dissolution in 25% HSA (human serum albumin) solution, and passage through an anion exchange column to remove inorganic ^{123}I. Up to 80% of the ^{123}I was incorporated in these products by this method.

Other substances recently reported to have been labelled with ^{123}I by methods similar to the foregoing are bromsulphophthalein (8),[119] Rose Bengal (9),[120] and two iodinated benzoic acid derivatives[121] marketed as contrast agents ['Conray' (10) and 'Hypaque' (11)] for use in diagnostic radiology. [^{123}I]HSA has also been

[116] M. D. Short, H. I. Glass, G. D. Chisholm, P. Vernon, and D. J. Silvester, *Brit. J. Radiol.*, 1973, **46**, 289.
[117] M. L. Thakur, B. M. Chauser, and R. F. Hudson, *Internat. J. Appl. Radiation Isotopes*, 1975, **26**, 319.
[118] G. D. Robinson, jun. and A. W. Lee, *J. Nuclear Medicine*, 1975, **16**, 17.
[119] M. L. Goris, *J. Nuclear Medicine*, 1973, **14**, 820.
[120] A. N. Serafini, W. M. Smoak, H. B. Hupf, J. E. Beaver, J. Holder, and A. J. Gilson, *J. Nuclear Medicine*, 1975, **16**, 629.
[121] M. L. Thakur, *Internat. J. Appl. Radiation Isotopes*, 1974, **25**, 576.

prepared[116] by an electrolytic method, which, unlike harsher methods, is reported not to denature the protein during labelling.

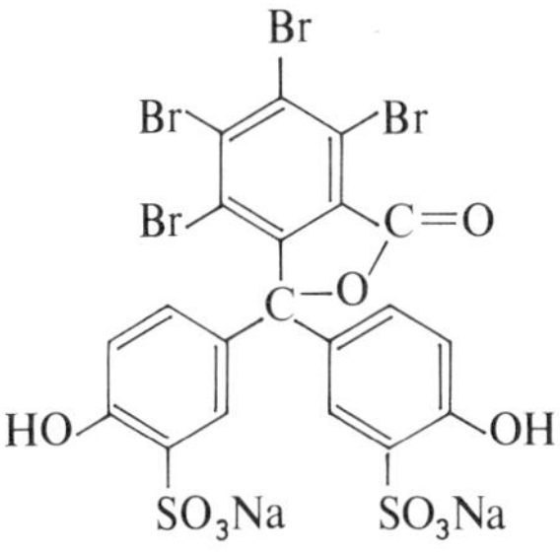

(8)

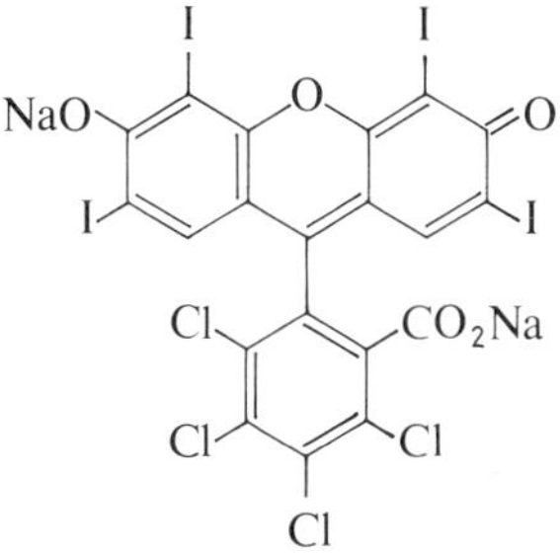

(9)

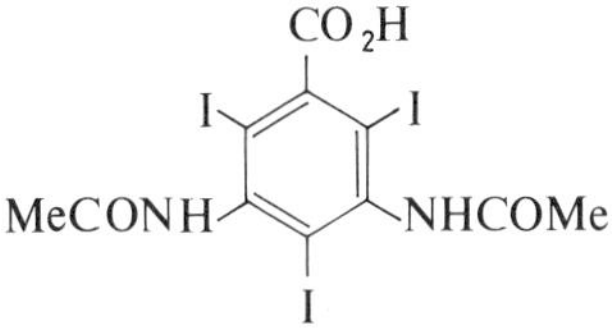

(10)

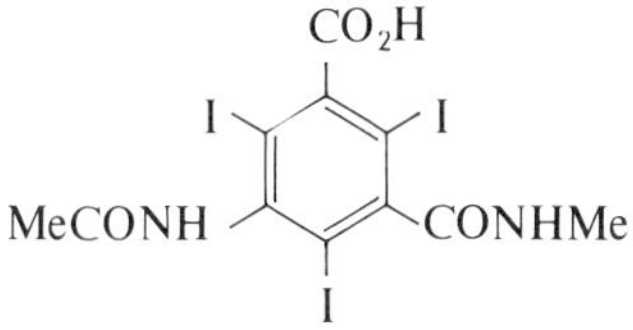

(11)

The decay of ^{123}Xe (28% by positron emission, 72% by electron capture leading to X-ray and Auger electron emission) results in the initial formation of highly charged ^{123}I species with quite low kinetic energy, and if this occurs in the proximity of organic molecules, labelling reactions may result. In discussing the principles of this process, to which the name 'excitation labelling' has been given, Lambrecht and Wolf[104] stress that it differs significantly from recoil, or hot-atom labelling, in which the labelling process is affected chiefly by the kinetic energy of the reacting species. Radiation damage to the organic substrate is usually non-existent with excitation labelling, except at very high activity levels.

The advantages claimed for this method include:

(i) carrier-free labelling can be attained (though without specificity of position),

(ii) high molecular weight, or complex, molecules can be labelled, where conventional synthetic methods might prove unsuccessful,

(iii) the procedure is fast and convenient, since the only chemical manipulation required is the repurification of the compound after the decay of the parent ^{123}Xe. An admitted disadvantage of the method is that in some cases the labelling yields are very low.

This approach has been used with some success to prepare [^{123}I]Indocyanine Green (ICG).[122] Crystalline ICG (2 mg) was placed in a Pyrex reaction vessel and carrier-free ^{123}Xe was adsorbed onto it as 77 K. After about 6 h, the remaining ^{123}Xe was evacuated, the dye was then dissolved and inorganic ^{123}I was separated from the labelled ICG. The specific activity of the product was 1—1.5 mCi mg^{-1}, equivalent to an incorporation of only about 10% of the available ^{123}I.

[122] A. N. Ansari, H. L. Atkins, R. M. Lambrecht, C. S. Redvanly, and A. P. Wolf, ref. 11, Vol. 1, p. 111.

^{123}I-Labelled iodoacetic, iodo-oleic, and iodopalmitic acids have all been prepared[123] by a slight variation of the foregoing method, but it gave very poor incorporation (2—3%) of ^{123}I in compounds such as deoxyuridine, L-tyrosine, and insulin. However, an extremely effective iodination reagent for the latter compounds was obtained when the ^{123}Xe was allowed to decay in a crystalline KIO$_3$ matrix. Upon dissolution in a dil. HCl solution of the organic substrate, immediate iodination was observed, and carrier-free mono-iodinated products containing 50—90% of the available ^{123}I were obtained.

Very high purity, carrier-free, [^{123}I]sodium iodide has been prepared[124] by rinsing the walls of a Pyrex vessel in which ^{123}Xe has been allowed to decay. However, if instead of rinsing out the ^{123}I, a suitable halogenated substrate is introduced which is then melted in the vessel, exchange labelling has been found to occur.[125] [^{123}I]-4-Iodophenylalanine, [^{123}I]-5- and [^{123}I]-6- iodotryptophan have been prepared in this manner.

The use of high pressure liquid chromatography to achieve rapid separation of carrier-free inorganic and organic compounds incorporating ^{123}I has been described.[126] Pretreated anion-exchange resin columns were effective for the separation of ^{123}I$^-$ and ^{123}IO$_3^-$ whilst ^{123}I-labelled model compounds (5-iododeoxyuridine and 5-iodouracil) could be separated from their uniodinated analogues on anion-exchange resin or silica columns.

9 Astatine-211

Radionuclides which decay by α-particle emission, if they can be incorporated into suitable biomolecules, are potentially of great interest in radiobiology and radiotherapy, because of the highly localized and intense dose of radiation they can deliver. Astatine, the heaviest of the halogens, and in particular ^{211}At ($t_{\frac{1}{2}} = 7.2$ h) appears to be especially suited to this role.

It has been noted[104] that the earlier literature appears to contain more reviews than original papers on astatine. Recently, however, two methods of preparing ^{211}At have been described. Both utilize the ^{209}Bi(α,2n) reaction, for which the Q-value is 22 MeV, and targets of bismuth metal, but differ in the method of recovery. Lindner et al.[22] melt the irradiated Bi in a stream of N$_2$ which sweeps the ^{211}At into a saline–sulphite trap. This appears to be a slow (up to 4 h) process, and considerable care has to be exercised if the product is to be recovered quantitatively.

Neirinckx and Smit[127] prefer wet chemistry and claim that their method gives consistently reproducible results. After dissolving the irradiated Bi in conc. HNO$_3$, ^{211}At is extracted from HCl–HNO$_3$ solution into di-isopropyl ether, to be back-extracted into NH$_2$OH,HCl solution. Destruction of hydroxylamine by H$_2$O$_2$ results in a dilute HCl solution of ^{211}At.

The preparation of ^{211}At-labelled proteins has been investigated[128] using keyhole

---

[123] M. El Garhy and G. Stöcklin, *Radiochem. Radioanalyt. Letters*, 1974, **18**, 281.
[124] R. M. Lambrecht, E. Norton, and A. P. Wolf, *J. Nuclear Medicine*, 1973, **14**, 269.
[125] R. M. Lambrecht, H. Atkins, H. Elias, J. S. Fowler, S. S. Lin, and A. P. Wolf, *J. Nuclear Medicine*, 1974, **15**, 863.
[126] K. Rossler, W. Tornau, and G. Stöcklin, *J. Radioanalyt. Chem.*, 1974, **21**, 199.
[127] R. D. Neirinckx and J. A. Smit, *Analyt. Chim. Acta*, 1973, **63**, 201.
[128] C. Aaij, R. J. M. Tschroots, L. Lindner, and T. E. W. Feltkemp, *Internat. J. Appl. Radiation Isotopes*, 1974, **26**, 25.

Preparation of Radiopharmaceuticals and Labelled Compounds 97

limpet haemocyanin(KLH) as a model protein, three methods of labelling being compared. The use of chloramine-T as oxidant resulted in only a very small fraction of the ^{211}At being bound to the protein. Electrolytic oxidation was more successful, giving up to 30% incorporation of ^{211}At after 30 min electrolysis at 1 V. However, attempts to improve the incorporation by the use of high potential differences (4—5 V) resulted in denaturing the protein. Most successful was the use of micromolar concentrations of H_2O_2 as the oxidant, when up to 40% of the ^{211}At could be bound to the protein. This was further improved to 60% by the addition of a small amount of KI with the H_2O_2. No evidence was found of the protein becoming denatured by this technique. The mechanism of this labelling procedure is not clear.

Labelling of surface membrane cellular antigens in baboon lymphocytes with ^{211}At by means of electrolytic oxidation has also been reported[129] but again the mechanism has not been elucidated and the labelled species remains uncharacterized.

As a preliminary requirement to a study of the mechanisms of astatination under different conditions, a number of relatively simple astato-compounds (*o*-, *m*-, and *p*- astatochloro-, -astatobromo-, and -astatoiodo-benzene) have been prepared[130] by decomposition of the corresponding diazonium salts in the presence of ^{211}At, analogous to the Sandmeyer reaction. The use of radio-gas-chromatography to identify these products has also been described in detail.

The separation of ^{211}At- labelled deoxyuridine and uracil by high-pressure liquid chromatography has been reported.[126]

10 Other Radionuclides

Readers who are already familiar with this subject may be surprised by the brevity of the remainder of this review, which is attributable to the dearth of professional chemists as yet working in this field. Unfortunately, whilst there are numerous reports of new radiopharmaceuticals, mostly, but not exclusively, in the medical literature, relatively few of these have been subjected to anything more than rudimentary chemical examination to establish their identity and stability, and for this reason many have been excluded from mention.

Frequently the method of preparing such radiopharmaceuticals consists merely of mixing a carrier-free radionuclide with the compound it is desired to label (perhaps adjusting the pH or temperature of the mixture) and then 'testing' it by observing the distribution of the radioactivity in animals, or even in man. It is hardly surprising, when for example, even the original carrier–radionuclide preparation may vary significantly depending upon its source,[131] that the results of these experiments are often irreproducible.

Sometimes the addition of carrier is assumed to remove uncertainty about the nature of the product, though, as any chemist will know, this will not be the case unless complete isotopic exchange can be shown to have occurred. Furthermore, analysis by paper chromatography or t.l.c. using solvents in which the supposedly labelled compound is found to have an R_f value of zero, and 'unbound' radioactivity a higher R_f value, is again inadequate proof that labelling has occurred.

In addition to those radionuclides reviewed in previous sections, many more pos-

[129] R. D. Neirinckx, J. A. Myburgh, and J. A. Smit, ref. 6, Vol. 2, p. 171.
[130] G. J. Meyer, K. Rossler, and G. Stöcklin, *Radiochem. Radioanalyt. Letters*, 1975, **21**, 247.
[131] A. D. Waxman, D. Kawadam, W. Wolf, and J. K. Siemsen, *Radiology*, 1975, **117**, 647.

Table 6

Radionuclide	Half-life	Principal Radiation/keV
^{67}Ga	78.3 h	γ(93,185,299)
^{68}Ga	68.3 min	$\beta^+:\gamma$ (1077)
^{99m}Tc	6.0 h	γ:(141)
^{111}In	2.83 d	γ:(245,171)
^{113m}In	99.5 min	γ:(392)
^{203}Pb	53.1 h	γ:(279,401)

sess physical characteristics (appropriate half-life and radiation) and chemical properties which make them potentially attractive for labelling purposes. However only a relatively small number have been much exploited so far. These are listed together with their physical characteristics in Table 6.

Gallium-67 and -68.—It is possible to produce ^{67}Ga in a cyclotron by several alternative nuclear reactions, induced by proton, deuteron, ^{3}He-, or ^{4}He- particles and using zinc or copper targets. A comparative review of early work, together with some original measurements of ^{67}Ga yields from Zn(d,xn), Zn(^{4}He, pxn), and ^{65}Cu(He, 2n) reactions, using 8 MeV deuteron- or 21 MeV ^{4}He-beams, has been published.[132] The authors found the last reaction gave the best of the yields which they measured [77.5 μCi (μAh)$^{-1}$] though this has to be compared with reported yields of 160 μCi (μAh)$^{-1}$ for the same reaction using a 30 Kev ^{4}He beam, or 400 μCi (μAh)$^{-1}$ for the Zn(p,xn) route using 23 MeV protons.

The yield of ^{67}Ga obtained by deuteron bombardment of natural zinc targets has been measured as a function of deuteron energy up to 16 MeV and found to peak at about 9 MeV.[133] From the resulting curve, the authors calculated a thick target yield of 352 μCi (μAh)$^{-1}$ and in practice 275 μCi (μAh)$^{-1}$ was achieved. Broadly similar yields [205—255 μCi (μAh)$^{-1}$] have been reported[134] using deuterons of 14 MeV maximum energy.

A group at Oak Ridge National Laboratory, the fount of enriched isotopes, reports[135] on the large-scale production of ^{67}Ga from the ^{68}Zn(p,2n) reaction, using enriched ^{68}Zn cyclotron targets. Target preparation requires considerable care to be taken with such expensive material, and the electroplating technique employed is described in some detail. The ^{67}Ga yield obtained by these authors was typically 3.5 Ci from an 8 h 22 MeV proton bombardment at 350 μA, which may be expressed as 1.6 mCi (μAh)$^{-1}$. As in the previously mentioned papers, the ^{67}Ga is recovered by solvent extraction following acid dissolution of the target. The method of recovering the ^{68}Zn for further use is also described, and $\leqslant 5\%$ loss per cycle is claimed.

The use of alumina adsorption columns, instead of solvent extraction, to effect recovery of ^{67}Ga from zinc or copper cyclotron targets has been carefully investigated.[136] By the recommended procedure, up to 65% of the ^{67}Ga could be obtained,

[132] F. Helus and W. Maier-Borst, ref. 6, Vol. 1, p. 317.
[133] J. Steyn and B. R. Meyer, *Internat. J. Appl. Radiation Isotopes*, 1973, **24**, 369.
[134] M. Vlatkovic, G. Paic, S. Kaucic, and B. Vekic, *Internat. J. Appl. Radiation Isotopes*, 1975, **26**, 377.
[135] L. Brown, A. P. Callahan, M. R. Skidmore, and T. B. Wilson, *Internat. J. Applied Radiation Isotopes*, 1973, **24**, 651.
[136] P. Kopecky and B. Mudrova, *Internat. J. Applied. Radiation Isotopes*, 1975, **26**, 323.

free from detectable contamination by other elements with the exception of aluminium.

Gallium-68 decays (like ^{11}C, ^{13}N, ^{15}O, and ^{18}F) entirely by β^+ emission, and thus, by virtue of the 180° correlation between the two subsequent annihilation radiation quanta, offers the possibility of detection with very high spatial resolution. This radionuclide is most conveniently obtained from generators, consisting of alumina columns loaded with the 375 d parent ^{68}Ge, which are now available from commercial sources. The daughter ^{68}Ga is normally eluted from such generators by a dilute edta solution; however, this forms a very strong complex which has to be broken down before attempts can be made to incorporate ^{68}Ga into other compounds.

One method of overcoming this problem reported[137] is to prepare a generator from which ^{68}Ga may be eluted with 0.2 M HCl. In an alternative approach[138] the ^{68}Ga–edta solution is made 6 M with respect to HCl, and passed through an anion-exchange resin column. This retains the ^{68}Ga but can be washed entirely free from edta by further 6 M HCl. Subsequently, *ca.* 80% of the ^{68}Ga is recovered by eluting the column with 1 cm^3 H$_2$O. The resulting solution is titrated to pH 2.3 with NaOH using phenol red as indicator. Alternative complexing agents can then be added (citrate and ATP are discussed) before the pH is further adjusted to 7. Methods of analysing the products by paper chromatography are also described.

Bleomycin (12) a; R = NH(CH$_2$)$_3$S$^+$Me$_2$X$^-$ (Bleomycin A)
b; R = NH(CH$_2$)$_4$NHCNH$_2$ (Bleomycin B)
              ‖
              NH
c; R = OH (Bleomycinic Acid)

[137] P. Kopecky and B. Mudrova, *Internat. J. Applied. Radiation Isotopes*, 1974, **25**, 263.
[138] J. Hnatowich, *J. Nuclear Medicine*, 1975, **16**, 764.

Gallium-67 citrate has been widely employed in medicine for the detection of tumours.[139,140] Investigations of the mechanism by which the radionuclide accumulates in tumour-cells have so far proved inconclusive, though there may be some relationship between ^{67}Ga uptake and the rate of DNA synthesis.[141] The cytotoxic drug Bleomycin (12),[141a] which is a mixture of related peptides with metal chelating properties, has also been labelled with ^{67}Ga.[142]

Indium-111 and -113m.—Four methods of preparing ^{111}In in a small cyclotron have been compared.[143] The nuclear reactions employed, yields of ^{111}In and radionuclidic contaminants are shown in Table 7. The authors point out that, whilst contamination could be substantially reduced by the use of isotopically enriched targets, this advantage would be off-set by the extra cost and time involved in target recovery. For regular production, proton bombardment of natural cadmium-foil targets is preferred.

Table 7 Yields of indium-111[a]

Desired reaction	Target thickness/ mm	Incident beam energy/ MeV	Yield (μCi/μA h at EOB)	Contaminants at 48 h (% of ^{111}In activity)	
^{112}Cd(p,2n)^{111}In	0.51	22	1035	^{114m}In	0.5
				^{110m}In	0.3
				^{109}In	0.4
^{110}Cd(d,n)^{111}In	0.25	12	117	^{114m}In	5.7
				^{110m}In	0.2
				^{109}In	0.02
^{100}Ag(^{3}He^{2+},n)^{111}In	0.25	32	2	^{110m}In	21
				^{109}In	126
^{109}Ag(^{4}He^{2+},2n)^{111}In	0.064	24	55	^{110m}In	0.7
				^{109}In	0.6

[a] Data taken from ref. 143.

Recovery of carrier free ^{111}In from irradiated Cd or Ag targets by means of solvent-extraction procedures has also been investigated.[144] In the favoured procedure, the target is dissolved in HNO$_3$ and ^{111}In extracted into bis(2-ethylhexyl) hydrogen phosphate. Back extraction into 5 M HBr is followed by a second extraction into butyl acetate from which it is finally stripped by water or 0.01 M HCl, to give a carrier-free product in which less than 10^{-6} of the target element is present.

Indium-113m is normally obtained by elution from generators loaded with its parent ^{113}Sn($t_{\frac{1}{2}} = 118$ d). The specific activity of the latter is usually very low,

[139] S. M. Larson, M. S. Milder, and G. S. Johnston, ref. 7, Chapter 43, p. 413.
[140] G. B. Saha and P. A. Farrer, ref. 7, Chapter 44, p. 435.
[141] P. A. G. Hammersley and D. M. Taylor, ref. 7, Chapter 45, p. 447.
[141a] T. Takita, Y. Muraoka, T. Yoshioka, A. Fujii, K. Maeda, and H. Umezawa, *J. Antibiotics (Tokyo)*, 1972, **25**, 755.
[142] M. L. Thakur, *Internat. J. Appl. Radiation Isotopes*, 1973, **24**, 357.
[143] N. S. MacDonald, H. H. Neely, R. A. Wood, J. M. Takahashi, S. I. Wakakuwa, and R. L. Birdsall, *Internat. J. Applied Radiation Isotopes*, 1975, **26**, 631.
[144] V. I. Levin, M. D. Kozlova, A. B. Malinin, A. S. Sevastianova, and Z. M. Potapova, *Internat. J. Applied. Radiation. Isotopes*, 1974, **25**, 286.

because it is made by the $^{112}Sn(n,\gamma)$ reaction, and ^{112}Sn has only 0.96% abundance in natural tin. This means that special care has to be taken in generator construction: small columns are desirable to achieve high concentration of ^{113m}In in the eluate, but on the other hand they must be large enough to prevent the possibility of Sn break-through.

Commercial generators are available in which the ^{113}Sn is adsorbed on a column of hydrous zirconium oxide (HZO). Recently, such columns have been compared[145] with an alternative which uses silica gel as the adsorbent. ^{113m}In may be eluted from both using dilute HCl. Whereas HZO has the higher adsorption capacity for Sn, the latter is claimed to have better operating characteristics, giving higher recovery of ^{113m}In, less dependent on HCl concentration, and being more tolerant of sterilization by autoclaving.

Other authors[146] have noted that not insignificant quantities of zirconium may be found accompanying ^{113m}In eluted from HZO columns. It was shown that whilst this had no discernible effect on the labelling of albumin macroaggregates, it did influence their subsequent stability.

The solution chemistry of carrier-free indium, especially where relevant to radiopharmaceutical preparation, has been studied in some depth.[147] Of particular

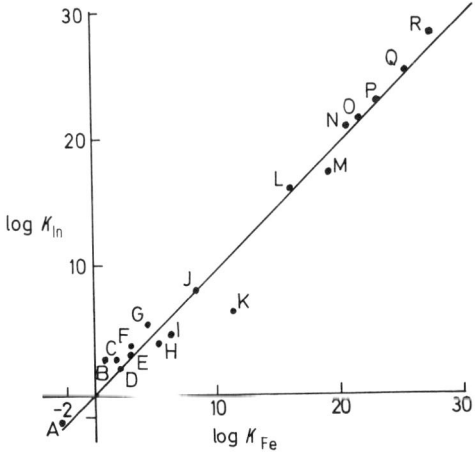

Figure 2 *Correlation of equilibrium constants for ferric and indium complexes.* A = OH^-; B = Cl^-; C = SCN^-; D = SO_4^-; E = *formate*; F = *acetate*; G = *oxalate*; H = F^-; I = *tartrate*; J = *acetylacetate*; K = *citrate*; L = *nitrilotriacetate*; M = *hedta*; N = $C_{12}H_{20}O_8N_2S$; O = $C_{11}H_{18}O_8N_2$; P = *eedta*; Q = *edta*; R = *dtpa*

(Reproduced by permission from 'Radiopharmaceuticals', ed. G. Subramanian, B. A. Rhodes, J. F. Cooper, and V. J. Sodd, Society of Nuclear Medicine, New York, 1975, p. 77)

[145] H. Arino and H. H. Kramer, *Internat. J. Appl. Radiation Isotopes*, 1974, **25**, 493.
[146] R. A. Caro, V. A. Ciscato, J. E. Ihlo, J. O. Nicolini, M. C. Palcos, and R. Radicella, *Internat. J. Applied Radiation Isotopes*, 1974, **25**, 501.
[147] M. J. Welch and T. J. Welch, ref. 7, Chapter 8, p. 73.

interest is the perhaps unexpected similarity between the stability constants of several In^{III} and Fe^{III} complexes, as shown in Figure 2. The implications of this similarity, and possible explanations for it, are discussed.

Simple complexes of both ^{111}In and ^{113m}In with edta and dtpa (diethylenetriaminepenta-acetic acid) have been used as radiopharmaceuticals, and their preparation, together with that of some other compounds, has been reviewed.[148,149] More recently, the preparation of four more ^{113m}In-complexes has been reported,[150] the complexing agents being:

> ethylenediaminetetra(methylene phosphonic) acid; edtmp
> hexamethylenediaminetetra(methylene phosphonic) acid; hmdtp
> diethylenetriaminepenta(methylene phosphonic) acid; dtpmp
> nitrolotris(methylene) phosphonate; ntmp

Complex formation was verified by a paper electrophoresis method, in which unbound indium remained at the origin, but the chelates migrated towards the anode.

The preparation of ^{111}In-bleomycin (*cf.* ^{67}Ga-bleomycin) was first reported in detail by Thakur[142] but has subsequently been discussed by others.[151,152] The complex is unaffected by heating or by exposure to Ca^{2+} ions.[152,153] Evidence has been presented,[152] however, to show that Cu^{2+} quickly displaces In^{3+} from the complex, and this could account for its reported instability *in vivo*.

Albumin and fibrinogen have been labelled with ^{111}In as examples of an intriguing new approach to the labelling of macromolecules by metal ions.[147,154] The basis of the method lies in the preparation of a so-called 'bifunctional chelate', azophenyl–edta (13):

$$\begin{array}{c} HO_2CCH_2 \\ HO_2CCH_2 \end{array}\!\!> NCHCH_2N <\!\!\begin{array}{c} CH_2CO_2H \\ CH_2CO_2H \end{array}$$

with a phenyl ring bearing N_2^+ attached to the central CH.

(13)

This compound can form a covalent link to suitable macromolecules (*e.g.* proteins) through the azo-group, whilst the edta group retains its ability to chelate metal ions.

Of further interest is the fact that, since its decay by electron capture is followed by the emission of two γ-rays in cascade,[111]In binding to macromolecules can be

---

[148] D. A. Goodwin, M. W. Sundberg, C. I. Diamanti, and C. F. Meares, ref. 7, Chapter 9, p. 80.
[149] J. Alvarez, ref. 7, Chapter 10, p. 102.
[150] G. Subramanian, J. G. McAfee, M. Rosentreich, and M. Coco, *J. Nuclear Medicine*, 1975, **16**, 1080.
[151] R. B. Grove, W. C. Eckelman, and R. C. Reba, *J. Nuclear Medicine*, 1973, **14**, 917.
[152] P. J. Robbins, E. B. Silberstein, and D. L. Fortman, *J. Nuclear Medicine*, 1974, **15**, 273.
[153] M. L. Thakur, M. V. Merrick, and S. W. Gunasekera, ref. 6, Vol. 2, p. 183.
[154] M. W. Sundberg, C. F. Meares, D. A. Goodwin, and C. I. Diamanti, *Nature*, 1974, **250**, 587.

Preparation of Radiopharmaceuticals and Labelled Compounds 103

studied experimentally by measuring the resulting perturbation of the angular correlation of the γ-rays. This method was used[154] to show, for example, that when 0.3 cm^3 of a 2% solution of human serum albumin, which had reacted with an equimolecular amount of azophenyl-edta, was added to 0.3 cm^3 of citrate buffer containing ^{111}InCl$_3$, binding was complete in less than 1 min.

Technetium-99m.—Technetium-99m is nowadays used in larger quantities than any other radionuclide for *in vivo* studies. This can be attributed first, to its almost ideal physical properties (Table 6) for this purpose; second, to its ability to adopt several stable valence states, so that it has been possible to incorporate it into a wide variety of compounds; and third, to its wide availability at low cost from commercial generator systems.

Several kinds of generator systems for ^{99m}Tc have been developed. The most widely used is the chromatographic type, consisting of an alumina column on which is adsorbed ^{99}Mo. This may be either of low specific activity, from the ^{98}Mo(n,γ) ^{99}Mo reaction, or carrier-free from the (n,f)^{99}Mo reaction. Advantages of generators prepared from fission-produced ^{99}Mo have been discussed[155] and, by virtue of the smaller chromatographic columns that can be employed, include higher ^{99m}Tc concentrations in the eluate.

Other kinds of generators include those which make use of solvent extraction, or of sublimation, to achieve separation of ^{99m}Tc from its parent, and all three systems have been compared[156,157] in detail by two groups of workers. More recently, careful studies have revealed impurities which could be significant, associated with ^{99m}Tc from extraction[158] and sublimation[159] generators, whilst attention has also been drawn[160] to the ^{99}Tc which lowers the specific activity of ^{99m}Tc from chromatographic generators in some circumstances.

The chemistry of technetium as applied to radiopharmaceuticals has been studied by none more thoroughly than by Richards and Steigman, who provide excellent summaries[161,162] of early work on this subject. Technetium-99m is initially recovered from all generator systems in its highest (+7) valence state as pertechnetate, and is frequently injected in this form. It is believed to be carried through the bloodstream weakly bound to albumin, which binds anions of all kinds by electrostatic forces. Similar forces may be responsible for exchange of ^{99m}Tc with Sephadex Gels, when the latter are used in the analysis of some weakly bound ^{99m}Tc complexes.[163]

Another widely used preparation is the so-called ^{99m}Tc–sulphur colloid (TcSC) in which, again, ^{99m}Tc is probably present in the +7 state. Steigman and Richards[161]

---

[155] H. Arino and H. H. Kramer, *Internat. J. Appl. Radiation Isotopes*, 1975, **26**, 301.
[156] R. E. Boyd, ref. 6, Vol. 1, p. 3.
[157] R. Constant, C. Fallais, R. Charlier, and P. Bievelez, ref. 6, Vol. 1, p. 27.
[158] M. W. Billinghurst, M. Groothedde, and R. Palser, *J. Nuclear Medicine*, 1974, **15**, 266.
[159] L. G. Colombetti, V. Husak, and V. Dvorak, *Internat. J. Appl. Radiation Isotopes*, 1974, **25**, 35.
[160] M. L. Lamson, A. S. Kirschner, C. E. Hotte, E. L. Lipsitz, and R. D. Ice, *J. Nuclear Medicine*, 1975, **16**, p. 639.
[161] J. Steigman and P. Richards, ref. 8, p. 269.
[162] P. Richards and J. Steigman, ref. 7, Chapter 3, p. 23.
[163] J. Steigman, H. P. Williams, P. Richards, E. Lebowitz, and G. Meinken, *J. Nuclear Medicine*, 1974, **15**, 318.

point out that there is no actual proof of this, but that in aqueous solution at macroscopic concentrations, carrier [^{99}Tc]pertechnetate will react with H_2S to form the insoluble Tc_2S_7.

Elsewhere[164] attention has been drawn to the importance of excluding nitrate from chromatographic ^{99m}Tc generators, especially if TcSC is to be prepared. This is because NO_3^- may be reduced to NO_2^- by γ-radiation resulting from the decay of ^{99}Mo on the column, which in turn may partially reduce TcO_4^- and lead to contamination of the TcSC by ^{99m}Tc in lower valence states.

TcSC has been used to label peripheral blood leukocytes by the process of phagocytosis during *in vitro* incubation.[165] The labelled cells must be isolated from whole blood with great care, so as to preserve their viability.

Many other substances can apparently be firmly labelled by ^{99m}Tc which has been reduced to lower valence states. The process of reduction by stannous chloride, which is widely employed for this purpose, has been investigated[166] using ^{99}Tc carrier. It was found that TcO_4^- is reduced to the +4 state in a citrate buffer at pH 7, to the +3 state in a dtpa buffer at pH 4, and possibly to the +4 state in HCl. Reduction can also be achieved by electrolysis, often carried out in dilute HCl solutions using a zirconium anode, when the nature of the reduced species remains uncertain.[167] It has, however, been shown that the existence of a zirconyl pertechnetate inner complex is extremely unlikely.

Because similar results were obtained in tissue distribution studies of two tin-reduced technetium complexes (gluconate and ehdp) when either carrier-free ^{99m}Tc or carrier ^{99}Tc were used,[168] it was concluded that reliable insight into the behaviour of ^{99m}Tc-radiopharmaceuticals can be gained from classical kinetic and thermodynamic work on ^{99}Tc adducts.

Problems associated with the labelling of proteins in general,[169] and albumin in particular[170] have been reviewed. All workers have found that ^{99m}Tc must be reduced by some means from the +7 state before firmly bound labelling can be achieved but the chemistry of the labelling process is still not understood. Studies using $FeCl_3$ and ascorbic acid to effect the reduction have led one group[171] to suggest that a hydrated ^{99m}Tc oxide may be implicated in the labelling of albumin. Another group,[172] working on the same system produce evidence which stresses the importance of carrying out the labelling procedure under a nitrogen atmosphere, to avoid the complications which can otherwise arise due to atmospheric oxidation.

Particles consisting of aggregates of ^{99m}Tc-labelled albumin molecules have proved useful diagnostic agents, and methods of preparing these have been des-

---

[164] A. G. Dumortier, O. M. Jeghers, P. Decostre, and C. J. Fallais, *Internat. J. Appl. Radiation Isotopes*, 1974, **25**, 189.
[165] D. English and R. A. Burton, *J. Nuclear Medicine*, 1975, **16**, 5.
[166] J. Steigman, G. Meinken, and P. Richards, *Internat. J. Appl. Radiation Isotopes*, 1975, **26**, 601.
[167] J. Steigman, W. C. Eckelman, G. Meinken, H. S. Isaacs, and P. Richards, *J. Nuclear Medicine*, 1974, **15**, 75.
[168] P. Hambright, J. McRae, P. E. Valk, A. J. Bearden, and B. A. Shipley, *J. Nuclear Medicine*, 1975, **16**, p. 478.
[169] M. S. Lin, ref. 7, Chapter 4, p. 36.
[170] B. A. Rhodes, ref. 8, p. 281.
[171] I. Zolle, L. Oniciu, and R. Hofer, *Internat. J. Appl. Radiation Isotopes*, 1973, **24**, 621.
[172] A. Yokoyama, G. Kominami, S. Harada, and H. Tanaka, *Internat. J. Appl. Radiation Isotopes*, 1975, **26**, 291.

cribed.[173,174] Other proteins to have been labelled include fibinogen[175] and streptokinase.[176]

Besides the leukocytes already mentioned,[165] other blood cells may be directly labelled with ^{99m}Tc, using using methods which have been reviewed by Eckelman and his colleagues.[177] Once again, it is advisable to reduce the ^{99m}Tc from the +7 state to achieve efficient and effective labelling, but once again the mechanism has not been elucidated. A method for labelling erythrocytes, using $SnCl_2$ as the reducing agent, has been described in detail.[178]

Bleomycin (cf. ^{67}Ga and ^{111}In) has also been labelled by $SnCl_2$-reduced ^{99m}Tc[179] as have tetracycline and related compounds,[180,181] dextrose,[182] and several phosphates[183] and phosphonates.[183,184]

Reduction of $^{99m}TcO_4$ can be effected by penicillamine, (14), or by related compounds.

$$\begin{array}{cc} SH & NH_2 \\ | & | \\ Me_2C - CH & CO_2H \end{array}$$

(14)

These are also powerful chelating agents, and capable of forming strong complexes with the reduced ^{99m}Tc. Particular interest has been shown in the compound formed when a ^{99m}Tc-penicillamine complex reacts with acetazolamine.[185]

Lead-203.—Early methods of producing ^{203}Pb have been summarized by Merrill *et al.*[186] All methods employed thallium as the target material, and the yields varied according to the particle beams used, as shown in Table 8.

The proceedings of a lead symposium[187] held at the Royal Postgraduate Medical School, London, in 1974 include a detailed description[188] of the preparation of ^{203}Pb using the $^{203}Tl(d,2n)$ reaction. The 15 MeV deuteron beam used gave yields of about 75 μCi $(\mu Ah)^{-1}$ (12 mCi from a 4 h bombardment) after chemical separation of the product. The latter, although nominally 'carrier-free' occasionally contained as much as 25 μg of lead which was present as an impurity in the target material.

[173] P. J. Robbins, D. L. Fortman, and J. T. Lewis, *Internat. J. Appl. Radiation Isotopes*, 1973, **24**, 481.
[174] D. B. Yeats, A. Warbick, and N. Aspin, *Internat. J. Appl. Radiation Isotopes*, 1974, **25**, 578.
[175] D. W. Wong and F. S. Mishkin, *J. Nuclear Medicine*, 1975, **16**, 343.
[176] B. R. R. Persson and V. Kempi, *J. Nuclear Medicine*, 1975, **16**, 474.
[177] W. C. Eckelman, T. D. Smith, and P. Richards, ref. 7, Chapter 5, p. 49.
[178] P. H. Smith, *Internat. J. Appl. Radiation Isotopes*, 1974, **25**, 137.
[179] M. S. Lin, D. A. Goodwin, and S. L. Kruse, *J. Nuclear Medicine*, 1974, **15**, 338.
[180] M. K. Dewanjee, C. Fliegal, S. Treves, and M. A. Davies, *J. Nuclear Medicine*, 1974, **15**, 176.
[181] C. D. Robinson, jun., and D. J. Battaglia, *Internat. J. Appl. Radiation Isotopes*, 1975, **26**, 147.
[182] J. Alvarez, C. Arriaga, and R. Maass, *Internat. J. Appl. Radiation Isotopes*, 1974, **25**, 283.
[183] G. Subramanian, J. G. McAfee, R. J. Blair, and F. D. Thomas, ref. 7, Chapter 34, p. 319.
[184] F. P. Catranovo, K. A. McKusick, M. S. Potsaid, D. Dolphin, T. L. Loewenthal, and R. J. Callahan, ref. 7, Chapter 7, p. 63.
[185] M. Tubis, G. J. Krishnamurthy, J. S. Endow, and W. H. Bland, ref. 7, Chapter 6, p. 55.
[186] J. C. Merrill, R. M. Lambrecht, and A. P. Wolf, *Internat. J. Appl. Radiation Isotopes*, 1973, **24**, 701.
[187] *Postgraduate Medical J.*, 1975, **51**, No. 601.
[188] P. L. Horlock, M. L. Thakur, and I. A. Watson, ref. 187, p. 751.

Table 8 Radionuclidic yields of cyclotron-produced lead-203[a]

Nuclear reaction	Particle energy/MeV	Radionuclide yield[b] at EOB ($\mu Ci/\mu Ah$)
$^{203}Tl(p,n)^{203}Pb$	12.5	67
$^{203}Tl(p,n)^{203}Pb$	16.2	50
$^{203}Tl(p,n)^{203}Pb$	20	70
$^{203}Tl(d,2n)^{203}Pb$	22.7	24.1
$^{203}Tl(^3He,3n)^{203}Bi \rightarrow ^{203}Pb$	29	7.5
$^{203}Tl(^4He,4n)^{203}Bi \rightarrow ^{203}Pb$	46	2

[a] Data from ref. 186.
[b] Yields measured after chemical separation, but corrected for decay since end of bombardment (EOB).

Production by means of Tl(p,xn) reactions with proton energies up to 45 MeV has also been studied,[189] and for practical purposes the chosen proton energy is 26 MeV, when, with a typical beam current of 5 μA, yields of about 10 mCi were claimed from bombardments of about 1 hour duration.

Lead-203 is a useful tracer in a variety of medical and environmental studies.[186,187] Its incorporation into tetraethyl-lead, for some of these studies, has been described.[189,190]

Table 9

Radionuclide	Half-life	Principal Radiation (keV)	Refs.
^{24}Na	15.0 h	β^-; γ (2754, 1369)	191
^{27}Mg	9.5 min	β^-; γ (844, 1014)	192
^{28}Mg	21.1 h	β^-; γ (31, 1342)	193
^{38}S	2.8 h	β^-; γ (1942)	194, 195
^{38}K	7.7 min	β^+; γ (2167)	196
^{43}K	22.2 h	β^-; γ (373, 618)	197
^{47}Sc	3.4 d	β^-; γ (159)	198
^{52}Fe	8.2 h	β^+; γ (169)	199
^{73}Se	7.1 h	β^+; γ (361)	200
^{81m}Kr	13 s	γ (191)	201—203
^{81}Rb	4.6 h	β^+; γ (448)	202—204
^{82}Rb	1.3 min	β^+; γ (776)	205
^{123}Xe	2.1 h	β^+; γ (149)	114, 206
^{125}Xe	16.8 h	γ (188, 243)	207
^{127}Cs	6.3 h	β^+; γ (411)	208
^{129}Cs	32.1 h	γ (372)	208
^{134m}Cs	2.9 h	γ (127)	209
^{137m}Ba	2.6 min	γ (662)	210
^{195m}Pt	4.0 d	γ (99, 130)	211
^{197m}Hg	23.8 h	γ (134)	212
^{201}Tl	73.5 h	γ (167, 135)	213, 214
^{206}Bi	6.2 d	β^+; γ (803, ...)	215

[189] F. Girardi, L. Goetz, E. Sabbione, E. Marafante, M. Merlini, E. Acerbi, C. Birattari, M. Castiglioni, and F. Resmini, *Internat. J. Appl. Radiation Isotopes*, 1975, **26**, 267.
[190] A. C. Chamberlain, W. S. Clough, M. J. Heard, D. Newton, A. N. B. Stott, and A. C. Wells, ref. 187, p. 790.
[191] Z. B. Alfassi, *Radiochem. Radioanalyt. Letters*, 1975, **22**, 87.

Miscellaneous Radionuclides.—Earlier reviews[5,14] have included references to several more radionuclides which have been used in simple radiopharmaceutical preparations. Table 9 lists those which have featured in more recent literature, together with their principle physical properties.

[92] Z. Alfassi and A. P. Kushelevsky, *Radiochem. Radioanalyt. Letters*, 1975, **21**, 87.
[93] T. Nozaki, M. Furukawa, S. Kume, and R. Seki, *Internat. J. Appl. Radiation Isotopes*, 1975, **26**, 17.
[94] L. Lindner, J. Visser, and H. Drost-Wildschut. *Internat. J. Appl. Radiation Isotopes*, 1973, **24**, 121.
[95] C. J. Leurs, L. N. Kremer, J. Boessma, and L. Lindner, *Internat. J. Appl. Radiation Isotopes*, 1975, **26**, 771.
[96] W. G. Myers, *J. Nuclear Medicine*, 1973, **14**, 359.
[97] F. C. Gray, C. M. Cole, G. M. Meaburn, and G. Brunhart, *J. Nuclear Medicine*, 1973, **14**, 931.
[98] T. Hara and B. R. Freed, *Internat. J. Appl. Radiation Isotopes*, 1973, **24**, 373.
[99] V. J. Sodd, K. L. Scholz, and J. W. Blue, *Medical Physics*, 1974, **1**, 25.
[200] T. Hara, S. Tilbury, B. R. Freed, H. Q. Woodard, and J. S. Laughlin, *Internat. J. Appl. Radiation Isotopes*, 1973, **24**, 377.
[201] L. G. Colombetti, L. W. Mayron, E. Kaplan, W. E. Barnes, A. M. Friedman, and J. E. Gindler, *J. Nuclear Medicine*, 1974, **15**, 868.
[202] P. V. Harper, B. Rich, K. A. Lathrop, and B. Mock, ref. 11, Vol. 2, p. 133.
[203] L. W. Mayron, E. Kaplan, A. M. Friedman, and J. E. Gindler, *Internat. J. Appl. Radiation Isotopes*, 1974, **25**, 237.
[204] B. Rich, N. Lembares, P. V. Harper, K. A. Lathrop, and F. Atkins, ref. 7, Chapter 18, p. 174.
[205] P. M. Grant, B. R. Erdal, and H. A. O'Brien, jun., *J. Nuclear Medicine*, 1975, **16**, 300.
[206] M. D. Loberg, M. E. Phelps, and M. J. Welch, *J. Nuclear Medicine*, 1973, **14**, 733.
[207] H. H. Lines, jun., N. F. Peek, G. L. DeNardo, and A.-L. Janshalt, *J. Nuclear Medicine*, 1975, **16**, 143.
[208] K. L. Scholz, V. J. Sodd, and J. W. Blue, *Internat. J. Appl. Radiation Isotopes*, 1974, **25**, 203.
[209] R. Chandra, P. Braunstein, F. Streuli, and J. Hirch, *J. Nuclear Medicine*, 1973, **14**, 243.
[210] H. Arino and H. H. Kramer, *Internat. J. Appl. Radiation Isotopes*, 1973, **24**, 197.
[211] H. S. P. Smith and D. M. Taylor, *J. Nuclear Medicine*, 1974, **15**, 349.
[212] T. Hara, M. Iio, M. Hirata, and T. Karasawa, *Internat. J. Appl. Radiation Isotopes*, 1973, **24**, 661.
[213] E. Lebowitz, M. W. Greene, R. Fairchild, and P. R. Bradley-Moore, *J. Nuclear Medicine*, 1975, **16**, 151.
[214] D. Comar, and C. Crouzel, *Radiochem. Radioanalyt. Letters*, 1975, **23**, 131.
[215] L. C. Brown and A. P. Callahan, *Internat. J. Appl. Radiation Isotopes*, 1975, **26**, 213.

4
Sample Preparation Procedures for Liquid Scintillation Counting

BY B. W. FOX

1 Introduction

The technique of liquid scintillation counting has been of outstanding value in a wide variety of applied and fundamental scientific disciplines since its introduction in the early fifties. The increasing sophistication of the instrumentation has resulted from the revolution in electronics following the invention of the transistor and more recently, of integrated circuit systems. The development of the instrument in its present form has been well reviewed by Rapkin.[1]

The physical chemistry of the scintillation process continues to be a source of considerable fundamental interest to photophysicists and photochemists alike. An interesting autobiographical review by Birks[2] has outlined the way in which research in this field has progressed. Some special applications of the scintillation process have been given by Horrocks,[3] in a review which also covers many of the newer ideas of the mechanisms of the fundamental processes involved.

Elsborg[4] has compared the liquid scintillation technique with gas proportional counting for the assay of tritium in biological materials. He suggests that the latter system may have certain advantages in non-homogeneous biological samples, since it is free from artefacts such as chemiluminescence, phosphorescence, and lack of solubility. Where an accurate assay of tritium content is required, there could be certain advantages in this technique, but improvements in liquid scintillation technology, in sample preparation, in instrument efficiency, and in automatic sample handling must make liquid scintillation counting the method of choice where large numbers of samples are involved.

Several useful introductory works[5—8] have been published during the period covered by this Report (1973—1974). A more detailed study by Kobayashi and Maudsley[9] on the biological applications of the liquid scintillation counting

[1] E. Rapkin, in 'Liquid Scintillation Counting', Vol. 2, ed. M. A. Crook, P. Johnson, and B. Scales, Heyden, London and New York, 1972, pp. 61–100.
[2] J. B. Birks, in 'Liquid Scintillation Counting, Recent Developments', ed. P. E. Stanley and B. A. Scoggins, Academic Press, New York and London, 1974, pp. 1—38.
[3] D. L. Horrocks, in 'Liquid Scintillation Counting', Vol. 3, ed. M. A. Crook and P. Johnson, Heyden, London and New York, 1974, pp. 3—20.
[4] L. Elsborg, *J. Nuclear Med.*, 1974, **15**, 115.
[5] A. Dyer, 'An Introduction to Liquid Scintillation Counting', Heyden, London and New York, 1974.
[6] L. W. Price, *Lab. Practice*, 1973, **22**, 110.
[7] L. W. Price, *Lab. Practice*, 1973, **22**, 181, 194.
[8] J. Klein, *Nuclear Instr. and Methods*, 1973, **112**, 117.
[9] Y. Kobayashi and D. V. Maudsley, 'Biological Applications of Liquid Scintillation Counting', Academic Press, New York, 1974, p. 196.

2 Basic Design of Instrumentation and Scintillation Vials

In response to an increasing demand for automation, sensitivity, and computational facilities, instrument manufacturers have developed instruments which exceed in precision and efficiency the still rather crude sample preparation procedures employed with them. The scope of the instrumentation has been extended, and liquid counting systems are now used in high-energy neutron studies,[11] for γ-discrimination as well as for the assay of α-emitters and many non-radioactive photon-emitting applications. A 6 l liquid scintillation counter has been built[12] for use with neutrons (70 MeV), for which it is reported to be 25% efficient. A large mineral-oil-based counter for the measurement of neutrons has also been described.[13] The scintillants in these cases are often loaded with an element like gadolinium in order to increase density, and a Monte Carlo simulation study of the capture and detection of neutrons by these large counters has been undertaken by Poitou and Signarbieux.[14]

Many modifications and suggestions for the use of scintillation counting vials have been made. The use of mini-vials, *i.e.* small glass or polythene vials[15,16] or microcentrifuge tubes[17] placed inside conventional vials, have often been described, especially in relation to radioimmunoassay procedures. Many of the vials are now commercially available. Even smaller volumes of 0.1 to 0.3 cm^3 total capacity have been assessed, by insertion of very small tubes supported on Kahn stoppers.[18] The use of small plastic bags have also been described, and increased efficiencies, especially in dual isotope assays have been claimed.[19] This is a useful point, since many molecular and biochemical procedures require an accurate assay of this type on such low volumes.

A vial has been described[20] for use in the continuous assay of $^{14}CO_2$ evolved *in situ*. A conventional scintillation vial may be used, lined with a cylinder of Whatman No. 40 paper impregnated with diphenyloxazole (PPO) and 1,4-di-2-(5-phenyloxazolyl)benzene POPOP, (the method of impregnation is described); the dry paper cylinder is wetted with 0.5 cm^3 of 1M NaOH and an inner vial containing the $^{14}CO_2$ generating system is inserted, and the outer vial is then capped. The $^{14}CO_2$ leaves the inner tube by a series of ports and is absorbed on the paper, and the increasing level of $^{14}CO_2$ is measured by the scintillations from the essentially heterogeneous system. Strictly speaking this is not a liquid scintillation counting procedure.

---

[10] R. Vaninbroukx and I. Stanef, *Nuclear Instr. and Methods*, 1973, **112**, 111.
[11] H. Ishikawa, *Nuclear Instr. and Methods*, 1973, **109**, 493.
[12] H. Nakamura, F. Reide, and T. Yuasa, *Nuclear Instr. and Methods*, 1973, **108**, 509.
[13] J. T. Dakin, *Nuclear Instr. and Methods*, 1974, **114**, 393.
[14] J. Poitou and C. Signarbieux, *Nuclear Instr. and Methods*, 1974, **114**, 113.
[15] J. R. Baur and R. W. Bovey, *Analyt. Biochem.*, 1974, **60**, 568.
[16] G. M. Connell and J. A. Linfoot, *Internat. J. Appl. Radiation and Isotopes*, 1973, **24**, 239.
[17] F. L. Schaffer and M. E. Soergel, *Appl. Microbiol.*, 1974, **28**, 280.
[18] K. D. Nearne and C. A. Homewood, *Analyt. Biochem.*, 1974, **57**, 623.
[19] R. Tykva, *Coll. Czech. Chem. Comm.*, 1973, **38**, 503.
[20] E. U. Buddemeyer, *Appl. Microbiol.*, 1974, **28**, 177.

A useful comparison of a range of vials currently used in liquid counting,[21a] and a report of a simple device for the pre-washing of vials prior to automated washing[21b] have appeared.

A modification has been described[22a] which allows for the measurement of fast pulse time intervals, rather than pulse height analysis. The modified instrument can accommodate up to 100 cm^3 of scintillant solution; this allows up to 20 g of water to be measured, at a level of accuracy of less than 10 tritium units.* Alternatively, tritiated benzene samples, up to 25 cm^3, may be measured at 53% efficiency with background rates of less than 1.75 c.p.m. Such low backgrounds require, of course, massive shielding as well as electronic guard systems. With the increase in tritium-producing nuclear reactor power facilities, there is an increased requirement for even more sensitive analytical systems for this isotope to monitor any possible environmental contamination.

An interesting review[22b] of the construction of the 4501 V4 photomultiplier (RCA Corporation) has appeared which also considers some possible future trends in photomultiplier development.

3 Solvents and Solutes in the Scintillator System

There has been little advance in the use of primary solvents in liquid scintillation counting and toluene, xylene, and dioxan–naphthalene have continued to be the solvents of choice. However, an exception which deserves further study is mentioned in a report of Krumbiegel and Schmidt.[23] They claim that by using a perdeuteriated solvent, such as a greater than 97% perdeuteriated toluene, the efficiency of counting tritium is almost 90% higher than with normal toluene.

Useful basic physicochemical data with regard to the radiative-transfer properties of the primary solvents, benzene, toluene, *o*-, *m*-, and *p*-xylene, and anisole have been recorded by Ishikawa and Takine.[24] Relative pulse-height data of diphenyloxazole solutions in a series of alkanes as solvents have been reported,[25] but in all cases, the values were less than 20% of the value in toluene.

Following the suggestion of Ashcroft,[26] the introduction of organic lead compounds into liquid scintillation cocktails as density-increasers to determine some γ-emitting isotopes, has been exploited, especially in radioimmunoassay techniques. Lead acetate solutions in water, shaken with ReadySolv VI (Beckman Instruments Ltd.) quickly produce an upper layer of scintillant suitable for these γ-measurements.[27]

There have been very few significant new primary solutes useful for scintillation counting reported during the period under review.

---

* The unit of such a small tritium level is a tritium unit (TU), one unit of which corresponds to a ratio of tritium atoms to hydrogen atoms of 10^{-18}:1. Alternatively, 100 TU's is equivalent to 6.5 d.p.m. tritium (g hydrogen)$^{-1}$.

[21] (*a*) K. Painter and M. J. Gezing, in ref. 3, p. 34; (*b*) N. S. Radin, *Analyt. Biochem.*, 1973, **55**, 637.
[22] (*a*) J. E. Noakes, M. P. Neary, and J. D. Spaulding, *Nuclear Instr. and Methods*, 1973, **109**, 177; (*b*) D. E. Persyk and T. T. Lewis, in ref. 3, pp. 21–27.
[23] P. Krumbiegel and H. Schmidt, U.S. P. 3 711 421, (1973).
[24] H. Ishikawa and M. Takine, *Nuclear Instr. and Methods*, 1973, **112**, 431.
[25] Y. Koike, *Nuclear Instr. and Methods*, 1973, **109**, 269.
[26] J. Ashcroft, *Internat. J. Appl. Radiation and Isotopes*, 1969, **20**, 555.
[27] E. Z. Helman and V. Spiehler, *Clin. Chem.*, 1974, **20**, 516.

However, an important advance, which may have far-reaching consequences with regard to the design of future primary solutes, is the recognition by Sharpe and Bransome[28] of the primary solute properties of the commercial detergent mixture BioSolv (BBS3) (Beckman Instruments Ltd.). This is a mixture of a nonionic alkyl phenyl ether and an anionic detergent. Its behaviour as a primary scintillator was recognized in its ability to count isotopes when in toluene alone, and was further confirmed by fluorescence excitation experiments, where it was shown to increase the fluorescence yield of toluene–PPO mixtures. This finding is not only of importance to the possible future design of primary solutes, but also points to the possible errors which could arise if quenched sealed standards are employed to compare quench data with detergent-containing mixtures.

The Raman spectra of diphenyl-2,5-furan (PPF), 2,5-diphenyl-1,3,4-oxadiazole (PPD), 2,5-di(4-biphenylyl)-1,3,4-oxadiazole (BBD), and *trans*-4,4'-diphenylstilbene (DPS) have been described.[29] The limiting factor in these studies was the solubility of the substances.

The problem of bringing the β-emitter into close thermodynamic contact with the electron-trapping system is simple where the sample is totally soluble in the primary solvent. Many samples employed in biochemical techniques require the assay of aqueous solutions often containing a complex mixture of a variety of salts and macromolecules. To overcome the incompatibility of the aqueous solution with the hydrocarbon solution of the primary solute, a third component which will effect phase unification with minimum loss of energy transfer is used, a so-called blender. In practice, such blenders are an important component of liquid scintillant mixtures and consist of primary alcohols, ethoxy-alcohols, or an amphiphilic cyclic ether, such as dioxan.

Most of the early reports on the application of blenders to the scintillant mixtures were derived empirically and designed only for particular systems. The wide variations in recommended mixtures are often due to a combination of local sample requirements, ease of access of solvents, and in many cases, just serendipity. Among the early reported formulations, some, *e.g.* those of Bray,[30] Butler,[31] and Kinard,[32] have been faithfully employed for a wide variety of samples, for which they may not be entirely suitable. Take, for example, the mixture suggested by Bray,[30] which consists of 60 g naphthalene, 4 g PPO, 0.2 g POPOP, 20 cm^3 ethylene glycol and 100 cm^3 methyl alcohol together with 1000 cm^3 of *p*-dioxan. Ethylene glycol was included originally in this mixture to allow the mixture to be employed at the lower emperature (*ca.* 7 °C) required by the earlier instruments, and to prevent the freezing of the dioxan. It does have some blending properties, especially in allowing a greater salt concentration at the 5—10 % sample volume.[33] However, the greater use of ambient temperature counting in modern instrumentation may mean a different composition would achieve the most efficient results and some of these older compositions need to be re-investigated systematically from this point of view.

[28] S. E. Sharpe and E. D. Bransome, jun., *Analyt. Biochem.*, 1973, **56**, 313.
[29] J.-M. Salmon and P. Viallet, *Bull. Soc. Chim. France I*, 1973, 2189.
[30] G. A. Bray, *Analyt. Biochem.*, 1960, **1**, 279.
[31] F. Butler, *Analyt. Biochem.*, 1961, **33**, 409.
[32] F. E. Kinard, *Rev. Sci. Instr.* 1957, **28**, 293.
[33] B. W. Fox, in 'Liquid Scintillation Counting', Vol. 4, ed. M. A. Crook and P. Johnson, Heyden, London and New York, (in press).

4 Homogeneity and Heterogeneity

There are two broad categories of sample preparation, monophasic (*i.e.* homogeneous) and multiphasic (heterogeneous). The monophasic systems are ideal insofar as the β-emitting sample is in close thermodynamic contact in the same phase as the scintillating primary solute. Such close contact demands that both sample and primary solute are present in true solution within the primary solvent, or a mixture of the primary solvent and a blender. The multiphase systems may be liquid–solid or liquid–liquid combinations in which the sample is in one phase and the scintillating solute is in the other phase. In many biochemical situations, the experiments only become possible if heterogeneous techniques are selected, since the actual sample preparation is usually much quicker and easier. The difficulties arise in assessing the degree of quenching in these systems, especially when the number of counts available is low and is statistically inadequate to allow an accurate assessment from the sample channels ratio quench-correction procedure (see Section 14).

Mueller[34] has suggested that there are certain conditions, even in a heterogeneous system, where the system may be regarded as homogeneous (operational homogeneity) and has suggested that a useful criterion of such conditions would be that the counting efficiency would be the same if any of the individual components of the system were labelled with an isotope.

In a heterogeneous system, the scintillant may be present in the solid or the liquid phase, and both systems find specific uses in radioassay applications. One of the earliest forms of scintillation counting of liquids was to suspend a solid scintillant in the form of beads in the solution to be assayed, but this method was too insensitive for general liquid scintillation counting, compared with the blended homogeneous systems and liquid emulsion systems. The main application of this form of heterogeneous counting is in the flow-through monitoring of liquids where complete recovery of the sample is a desirable feature and where the main requirement is the detection of the location of label in column or gaseous effluent from analytical procedures, rather than accurate measurement of the amount of radioactivity present.

The commonest heterogeneous counting system is where the β-emitter is in the solid phase and the liquid scintillator is in the liquid phase surrounding it. Measurements of samples as suspensions, or on solid surfaces such as filter discs and membranes, are examples of the application of this method.

Of considerable current interest however, is the liquid–liquid, two-phase system where the scintillant is in one phase and the sample is in the other. This form of heterogeneous system is the basis of the colloidal scintillation counting system.

5 Preprocessing for Homogeneous Liquid Scintillation Counting

The prime object of preprocessing for homogeneous scintillation counting is to ensure that the sample is present in the same phase as the primary solvent. This is usually achieved by combustion or degradative procedures which convert the toluene-insoluble macromolecular components into soluble, low molecular weight materials which are easily blended into the primary solvent, with alcohols or other

---

[34] E. B. Mueller, in ref. 3, p. 52.

amphiphilic solvents. Alternatively, the use of toluene-soluble complexes or lipophilic salts has frequently been employed, especially in the inorganic field.

An alternative method is to partition the labelled product to be measured into the organic, primary solvent phase and several examples of this type of preprocessing have been described, especially in relation to the measurement of enzyme activity. A good example of this procedure is in the assay of the enzyme, methyl transferase.[35] Further examples of this procedure are described in Section 16.

A comprehensive and useful survey of the preparation of homogeneous samples for liquid scintillation counting has been given by Gordon.[36]

6 Inorganic Ions

There have been several reports concerned with the analysis and monitoring of nuclear power station effluents as well as the level of some of these isotopes in the environment. One of the most active fields has been the simultaneous measurement of plutonium-241 β-emission and plutonium α-activity, since a knowledge of the level of each type of activity can give important information on the quality of the original 'burn up' which produced the contamination, as well as on the age of the source.

One example of this type of measurement is the co-precipitation of plutonium ions on $BaSO_4$ and extraction of the plutonium into di-(2-ethylhexyl)phosphoric acid, with which the plutonium forms a toluene-soluble complex salt.[37] It can then be extracted into a toluene-based scintillant for assay. The lower limit of detection by this technique is reported to be approximately 1 pCi. The samples of plutonium are often taken as smears from surfaces, and such smears have been measured in the past by a combination of internal proportional counting and α-scintillation counting. A method has been described[38] for an in-vial oxidation procedure using ammonium persulphate on the base of the vial, and the disc supported above the salt within finely divided silica. The whole is placed in a muffle furnace and after cooling toluene-based scintillant is added and the isotopes assayed as a suspension. High separation efficiencies are claimed by this procedure.

A rapid and accurate method for the determination of plutonium in urine has also been described[39] using ultrafiltration as a basic separation and concentration technique. Using sensitive liquid scintillation detection equipment, levels of plutonium of the order of 0.11 d.p.m. in a 24 h urine sample were detected. Methods for the assay of plutonium in air filters and in soil have also been described.[40]

The distribution of americium-241 in the rat has also been studied[41] and has found to de dependent not only on the dose given but also on the pH of the injection solution. Iron-55 also occurs as a low-level discharge of nuclear power stations, and this isotope is eventually concentrated in fish blood. The processing and purification of iron samples from this source has been described[42] leading to the formation of a

[35] L. Sankaran and B. M. Pogell, *Analyt. Biochem.*, 1973, **54**, 146.
[36] B. E. Gordon, in ref. 3, pp. 109—121.
[37] K. G. Darrell, G. C. M. Hammond, and J. F. C. Tyler, *Analyst*, 1973, **98**, 358.
[38] J. E. Eakins, in ref. 3, p. 287.
[39] G. N. Stradling, D. S. Popplewell, and G. J. Hahn, *Internat. J. Appl. Radiation and Isotopes*, 1974, **25**, 217.
[40] D. L. Bobowski, *Amer. Ind. Hygiene*, 1974, **35**, 333.
[41] A. Seidel, *Internat. J. Appl. Radiation and Isotopes*, 1973, **24**, 362.
[42] G. A. Sutton and B. H. Harvey, in ref. 3, pp. 279—286.

purified iron phosphate which is assayed in a water-accepting scintillant. Colloidal scintillation counting techniques using sodium xylene sulphonate and Triton N101 have also been recommended for the assay of environmental levels of ^{55}Fe.[43]

The use of dibutyl phosphate as an extracting agent for inorganic ions, is now well established in this field, and the method has been extended[44] to the analysis of mixtures of ^{95}Zr and ^{95}Nb. In this case the scintillant mixture employed was 0.4 g *p*-terphenyl and 0.1 g POPOP per litre of xylene. Alternatively, an aqueous solution of the ions was assayed in a mixture (v/v) of dioxan (75%), anisole (12.5%) and a solution of PPO and POPOP in 1,2-dimethoxyethane (12.5%).

Radium has been measured by trapping the radon produced onto a fine mesh silica gel, cooled in liquid nitrogen.[45] The silica gel is then assayed in a toluene-based scintillant mixture. The silica gel, purified grade, was ground to a mesh size of 14—30 inch^{-1} and dried at 105 °C for 2 d before use. The lowest level of radon that could be detected by this procedure was reported to be approximately 0.1 pCi. The radium activity is determined from this data using an 'in-growth factor' f (equal to $1-e^{-\lambda t}$) and E, the c.p.m. pCi^{-1} of radon expected.

$$\text{Radium activity} = \frac{c}{Ef} \text{ pCi (where } c = \text{c.p.m. obtained less background count rate)}$$

The special problems associated with the standardization of ^{87}Y and ^{67}Ga have been discussed[46] and it is pointed out that the apparent internal conversion electrons are associated with long-lived excited states which are a feature of the decay of these radionucleides. The efficiency of any assay using pulse-height selection methods will thus depend on a knowledge of the intensities of γ-emission and the efficiencies of the liquid scintillation counting system for detecting these emissions.

The assay of β- and γ-emissions in liquid scintillation counters has provided a means of assaying such isotopes as ^{46}Sc, ^{59}Fe, ^{60}Co, and ^{68}Rb quite efficiently[47] and very good agreement with γ-ray spectrometric results has been achieved.

7 Combustion Methods

For measurement of both tritium and carbon-14 in biological material, quantitative combustion in oxygen and absorption of the tritiated water and [^{14}C]carbon dioxide so formed into suitable scintillant mixtures is undoubtedly the most accurate method of determining the absolute level of activity present. Many devices have been described over the last decade and several commercial instruments are now available for this purpose.

Bomb combustion has been used for a variety of biological samples and with greater care in the construction of the bomb, yet larger sample sizes may be employed. In radiocarbon analysis, approximately 12 g of charcoal often needs to be combusted and dangerous 'dust explosions' may be experienced if incorrect pro-

---

[43] A. A. Moghissi, E. I. Whittaker, D. N. McNeils, and R. Lieberman, *Analyt. Chem.*, 1974, **46**, 1355.
[44] J. D. Ludwick, U.S. Atomic Energy Commission Report, HW 61728, 1959.
[45] K. G. Darrall, P. J. Richardson, and J. F. C. Tyler, *Analyst*, 1973, **98**, 610.
[46] J. Steyn, *Nuclear Instr. and Methods*, 1973, **112**, 137.
[47] H. Ishikawa and M. Takiue, *Nuclear Instr. and Methods*, 1973, **112**, 437.

cedures are used. The construction of an efficient bomb for this purpose and a detailed procedure for its use has been described.[48,49]

The alternative tube-furnace combustion method has been automated for repeated liquid sample preparation and the development of a commercial instrument based on this method has been described.[50]

The oxygen-flask method of assay is usually simpler to construct for general laboratory applications and many variations on this theme have been published. A successful commercial modification of the method has not been in use for a number of years. A simpler laboratory automated flask combustion system has now been described,[51] built from eight 500 cm^3 to 1000 cm^3 Erlenmeyer flasks. A focused beam from a projector lamp is used to initiate the ignition and following a 6 min cooling period, defined volumes of absorbing scintillants are introduced and aliquots removed for radioactive assay. An automated, in-vial, combustion technique has also been described with a view to commercial production.[52] Simple oxygen-flask procedures for faecal ash[53,54] have also been reported.

A useful comparison of locally constructed and commercially available flask combustion with tube-furnace techniques has been made[55] according to which the oxygen-flask method appears to be the most reliable procedure.

8 Solubilization methods

The incorporation of organic material into a non-polar or blended toluene-based scintillant, requires the conversion of any macromolecular species present into a degraded or complexed form which will be soluble in these solvent systems. Quaternary ammonium salts containing both aliphatic and aromatic hydrocarbons have been successfully employed in this field with a wide range of biological materials, even though the exact mechanism of complex formation with protein molecules is still unknown. Vaughan *et al.*[56] early recognized the value of the commercial germicidal agent, Hyamine 10X (Rohm and Haas, Inc.) 4-(1,1,3,3-tetramethyl-butyl-cresoxyethoxyethyl dimethyl benzyl ammonium hydroxide) for its ability to 'solubilize' amino-acids and proteins and to allow their dissolution subsequently in a toluene-based scintillant mixture. A number of other complex organic bases have since been used and appear as commercial formulations, *e.g.* BioSolv 1, BioSolv 3 (Beckmann Instruments), NCS, PCS (Nuclear Chicago-Searle), and Soluene 100 (Packard Instruments). Some formulations however also contain detergents, and thus convert the counting system into a heterogeneous colloidal system with its attendant quench-correction problems. Under many counting conditions used, especially with carbon-14, the system can usually be regarded as 'operationally homogeneous'.[34]

---

[48] V. R. Switsur, R. Burleigh, N. Meeks, and J. M. Cleland, *Internat. J. Appl. Radiation and Isotopes*, 1974, **25**, 113.
[49] R. Burleigh, in ref. 3, pp. 295—302.
[50] E. Rapkin, in ref. 3, pp. 132—149.
[51] R. Rauschenbach and H. Simon, in ref. 3, pp. 158—163.
[52] L. A. Wegner and H. Winkelmann, in ref. 3, pp. 150—157.
[53] T. Arnfred, K. Fugh, and L. Pedersen, *Scand. J. Gastroenterol.*, 1974, **9**, 325.
[54] J. G. Steytler, *J. Clin. Pathol.*, 1974, **27**, 844.
[55] R. G. Cooper, in ref. 3, pp. 164—174.
[56] M. Vaughan, D. Steinberg, and J. Logan, *Science*, 1957, **126**, 446.

Certain types of quaternary ammonium bases employed in commercial solubilizers have been shown to be incompatible with some primary solutes[57,58] such as butyl PBD 2[4'-tert-butylphenyl-5(4"-biphenylyl)-butyl]-1,3,4-oxadiazole owing to the development of excessive colour quenching. Certain other combinations, such as xylene, Triton X100, and BBOT [2,5-bis(5'-tert-butylbenzyloxazolyl)thiophen] have also been shown[59] to produce colour quenching but no mechanism has yet been proposed for the formation of colour in these systems.

Membrane filters based on cellulose acetate or mixed esters are frequently used in nucleic acid studies. In order to increase the efficiency of counting of these filters, tissue solubilizers have been used.[60] However, colour quenching is also a serious problem in these procedures and a study of the method of prevention of this colour has been undertaken.[61] It was found that colour quenching is produced when certain tissue solubilizers are used with the mixed ester type of filter [*e.g.* HAWG (Millipore Corporation)] but not with the cellulose acetate filters [*e.g.* EHWG (Millipore Corporation)].

The increasing number of commercially available solubilizers has presented the laboratory worker with a difficult problem of choice for any particular application. Attempts have been made to compare[62] the efficiencies of a number of different commercial solubilizers but a good systematic survey with a wide spectrum of sample, is still required. A useful comparison[63] of commercial solubilizers and procedures based on sodium hydroxide–methanol mixtures or perchloric acid suggests that little advantage exists in using the first although there is generally a considerable difference in the cost.

9 Cerenkov Light Measurement

The assay of the more energetic β-emitters by the measurement of Cerenkov emission is now well established and has found wide application in biomedical, inorganic, and plant physiology fields.[64,65] The method is applicable only to those β-emitting isotopes whose energy exceeds 0.263 MeV in water. The light emitted spans the spectral range from 300 to 700 nm, overlapping the sensitivity range of the photomultipliers used in the liquid scintillation spectrometer; however, being essentially one-photon events, greater efficiency is obtained by using only a single photomultiplier tube for the detection of the events rather than the coincidence system used in most commercial systems. In most commercial instruments the conversion from coincidence to single photomultiplier counting is a simple switching operation.

The measurement of Cerenkov emission has especially useful characteristics in sample preparation, since no scintillator system is required and the sample may therefore occupy the whole volume of the vial. This results in a considerable increase in the number of counts available to the counter, even though the absolute efficiency

[57] A. Dunn, *Internat. J. Appl. Radiation and Isotopes*, 1971, **22**, 212.
[58] K. Painter and M. J. Gezing, *Internat. J. Appl. Radiation and Isotopes*, 1973, **24**, 361.
[59] L. R. Wetter and J. Dyck, *Internat. J. Appl. Radiation and Isotopes*, 1973, **24**, 430.
[60] K. A. O. Ellem, *Biochem. Biophys. Acta*, 1967, **149**, 74.
[61] R. E. Johnsonbaugh, J. O. Kleiman, and J. Sode, *Analyt. Biochem.*, 1973, **54**, 490.
[62] G. W. Carter and K. Van Dyke, *Analyt. Biochem.*, 1973, **54**, 624.
[63] J. G. Dent and P. Johnson, in ref. 3, pp. 122—131.
[64] R. P. Parker and R. H. Elrick, in 'Liquid Scintillation Counting', ed. E. D. Bransome, jun., Grune and Stratton, New York and London, p. 110.
[65] K. Kisama, *Jap. Analyst*, 1974, **23**, 294.

may be much lower, *i.e.* it has a high merit value (percentage efficiency × percentage sample volume) the percentage sample volume in this case being 100. Provided the sample is optically clear and free from colour, relatively high counting efficiencies may be achieved with such isotopes as ^{32}P, ^{86}Rb, and ^{42}K. Two advantages are that the sample may be recovered following assay, and that impurity quenching plays little or no part. The presence of colour considerably decreases counting efficiency, but the lack of impurity quenching sensitivity allows strong bleaching agents to be used to remove the colour. A useful review of the modern applications of this technique was given by Parker.[66]

The measurement of small volumes of ^{32}P solutions by this technique presents severe centring problems within the vial and these have been examined.[67] With careful design of the vial cap and centring supports, a recommended method[68] uses a 12 × 57 mm polystyrene tube within conventional polyethylene vials. Higher efficiencies are reported to be obtained for volumes less than 2 cm³. The Shöniger flask combustion method has been used[69] to assay ^{32}P and ^{36}Cl in biological material by Cerenkov light measurement. The products of combustion in this case are dissolved directly in water and assayed without further processing. The direct assay of ^{32}P in intact plant roots using Cerenkov measurement in a liquid scintillation counter has also been described and the quenching due to the root structure itself has been estimated.[70]

The combined application of Cerenkov emission measurement and conventional liquid scintillation counting has been suggested as a successful procedure for the assay of mixtures of strontium-90 and strontium-89.[71]

10 Preprocessing for Heterogeneous Liquid Scintillation Counting

Many biochemical procedures result in the need to assay a β-emitter absorbed within or on the surface of a solid surface. For an accurate assessment of the β-emitter level it is necessary to elute the sample from the solid into a homogeneous scintillator solution. However, the main advantages of the measurement of β-emitters by heterogeneous techniques rest mainly in the increased ease of sample preparation. Samples absorbed onto thin layer surfaces, such as silica gel, cellulose, ion exchange, *etc.*, can usually be dried under i.r. lamps and assayed directly without further processing in non-polar liquid scintillator mixtures. If the absorbed material is water soluble, it is usually more efficient to remove the β-emitter by the use of polar-blended scintillant mixtures, and to assay as a homogeneous system. The usual trouble with such procedures is the partial solubilization of the labelled material which results in considerable variability and difficulty in determining the level of quenching that occurs. The same difficulties apply to labelled materials absorbed on chromatography paper and membrane strips following electrophoresis.

Heterogeneity occurs in most, if not all, scintillation-sample mixtures that contain detergent. However, the effect of detergent in bringing into close thermodynamic contact an aqueous solution containing the radioisotope, admixed with interfering

[66] R. P. Parker, in ref. 3, pp. 237—252.
[67] R. T. Hairland and L. L. Bieber, *Analyt. Biochem.*, 1970, **33**, 323.
[68] S. Mardh, *Analyt. Biochem.*, 1975, **63**, 1.
[69] A. R. Britt, Abstracts of the 167th A.C.S. National Meeting, 1974, NUCL 41.
[70] F. S. Chapin and D. F. Halleman, *Internat. J. Appl. Radiation and Isotopes*, 1974, **25**, 568.
[71] R. Randolph, *Internat. J. Appl. Radiation and Isotopes*, 1975, **26**, 9.

salts and macromolecules, and an efficient, immiscible, non-polar scintillant solution creates an unusually efficient mixture for the assay of such samples. The levels of detergent and the two phases are critically important to both the efficiency and the stability of the counting mixtures and these parameters have been studied (see Section 13). However, the resulting emulsion (or more correctly, colloid) system of radioassay has considerably extended the methods of sample preparation technique available, and several commercial, colloidal based mixtures have now become available.

Although not strictly a liquid scintillation system, the measurement of β-emitters by the use of solid scintillators is the basis of many continuous-flow monitoring techniques. However, there are also stream-splitting devices which allow an automated monitoring of a continuous system, such as the effluent of a gas or liquid chromatography system, using true liquid scintillation counting methods and these are reviewed below (Section 11).

11 Monitoring of Continuous-flow Systems

There are two basic methods by which a continuous monitoring of output material from gas or liquid chromatographic procedures may be undertaken. A solid scintillant placed within the stream of flowing gas or liquid may be used, or an aliquot of the stream can be removed by stream-splitting devices and assayed by liquid scintillation counting procedures. The advantage of the latter method is the several-fold increase of efficiency of the technique. Different methods have been reviewed for both gas and liquid chromatographic applications.[72,73]

A recent example[74] of the use of such flow monitors applied to liquid chromatography is in the automatic assay of cross-links in collagen samples reduced with sodium borohydride and fractionated on a column. The technique has been described[74] (but not illustrated) and consists of a proportionating pump (Technicon Instruments Ltd.) removing a continuous sample from the eluate at 0.16 cm^3 min.$^{-1}$ This is mixed with a scintillant solution flowing at 5 cm^3 min.$^{-1}$ The scintillant mixture used was 5 g butyl PBD, 0.5 g PBBO, 50 cm^3 BioSolv 3 (Beckmann Instruments Ltd.) per litre of toluene. The mixed sample/scintillant is assayed as discrete aliquots, using a commercial discrete-sampling flow-cell apparatus. After a 30 s count, the vial is automatically emptied. The advantage of this system is that it allows a sequential series of results to be recorded on paper tape for further data analysis.

The continuous measurement of a gas chromatograph analysis has also been described[73] and requires the removal of the carrier gas, and concentration of the products, coupled with a controlled mixing and measurement of the effluent.

12 Scintillant in the Liquid Phase

The assay of β-emitters on solid supports has gained wide usage in liquid scintillation counting, in the inorganic field and especially in biomedical applications. Glass fibre sheet has the useful property of allowing a film of a sample to be deposited on its surface, with no penetration of the support (except between fibres). The disc

[72] J. P. Verhassel, *Mess Technik*, 1973, **81**, 53.
[73] L. Schutte and E. B. Koenders, *J. Chromatog.*, 1973, **76**, 13.
[74] G. L. Mechanic, *Analyt. Biochem.*, 1974, **61**, 349.

(usually 24 mm diameter) can be dried under standardized conditions after application of the aqueous sample and added directly to a small volume of a non-polar toluene-based scintillant in a vial and assayed. Up to 26 such discs can be added to a single vial, and provided they are covered with scintillant, can be assayed stoichieometrically related to the disc number. The simple non-blended toluene scintillant usually employed in such procedures is a 4 g l^{-1} solution of PPO in toluene, so that maximum efficiency across a solid–liquid phase boundary can be maintained. In the deposition of the sample it is important to maintain a uniformly thin layer. However, although the main advantage of this technique is speed of sample preparation, the final counting conditions for a weak β-emitter like tritium, will however be essentially 2π. In many cases, careful and judicious use of solubilization techniques may be used which would allow for the considerably improved 4π counting of a homogeneous system offset only minimally by quenching.

An example of the use of glass-fibre discs is the analysis of ^{241}Am in urine.[75] The method has been extended[76] to the bioassay of ^{243}Am and ^{244}Cm, by utilizing the known high binding properties of the lanthanides and actinides towards glass surfaces. Indeed, this sample preparation technique especially for ^{244}Cm, requires special care to prevent the strong adsorption of the isotopes to other glass surfaces during preparation.

Variability in radioassays of monomeric and polymeric carbohydrates on paper chromatograms has also been studied.[77] It was shown that the addition of water to remove the carbohydrate prior to the addition of a xylene-based blended scintillant mixture [e.g. Aquasol, (New England Nuclear Corporation)] resulted in a three- to five-fold improvement in counting efficiency of both ^{3}H- and ^{14}C-labelled carbohydrates. For the extraction of lipids from thin layer surfaces, the addition of a few drops of methanol or ethanol to the scintillant has been recommended;[78] this reduces the level of self-adsorption of the lipid to the surface.

Cation exchange paper, such as DEAE (diethylaminoethyl cellulose) has been successfully employed as a solid support which selectively binds one or more labelled components under defined ionic conditions. Several enzyme assays have been described[79] based on this principle. Simple assays for the determination of methionine adenosyltransferase enzyme activity using phosphocellulose (P81, nominal capacity 18 Eq cm^{-2})[80] or carboxymethylcellulose[81] have been described.

It is often necessary to remove amino-acids and t-RNA from filter paper strips after radioactive analysis for further processing and a convenient method of undertaking this from Whatman No. 1 paper has been described.[82]

The special problems associated with the absorption and analysis of tritiated DNA on both glass fibre and nitrocellulose filters have been examined in detail.[83]

---

[75] J. D. Eakins and P. J. Gomm, *Health Phys.*, 1968, **14**, 461.
[76] F. E. H. Crawley, *Internat. J. Appl. Radiation and Isotopes*, 1975, **26**, 137.
[77] P. A. Sandford and P. R. Watson, *Analyt. Biochem.*, 1973, **56**, 443.
[78] J. A. Pyrovokakis, D. S. Harry, M. J. Martin, and N. McIntyre, *Clin. Chim. Acta*, 1974, **50**, 441—444.
[79] K. G. Oldham, Radiochemical Methods of Enzyme Assay, Review No. 9, 1968, The Radiochemical Centre, Amersham, England.
[80] R. M. McKenzie and R. K. Gholson, *Analyt. Biochem.*, 1973, **53**, 384.
[81] R. H. Wilson, *Biochem. J.*, 1970, **118**, 16.
[82] M. R. V. Murphy and H. Roux, *Analyt. Biochem.*, 1974, **58**, 89.
[83] B. K. Schrier and S. H. Wilson, *Analyt. Biochem.*, 1973, **56**, 196.

A number of different solubilization methods were assessed. Collection of high molecular weight DNA on glass fibre discs, then removal by solubilization in NCS [Nuclear Chicago Solubilizer (Searle)] followed by assay in a toluene-based scintillant was considered the most efficient method, especially when bovine serum albumin was used as a co-precipitant with the DNA. However, for single-stranded DNA of molecular weight $ca.$ 1.85×10^5 or lower, nitrocellulose filters were clearly superior. Additional bovine serum albumin did not contribute towards either the efficiency of collection or of radio-assay. Definite and useful recommendations were made as to the methods of filtration, radioassay, and for the reduction of blanks in the DNA polymerase assay system.[83]

A most significant advance in assaying tritium on solid supports appears to be the use of extruded expanded polystyrene foam discs 1/16th inch thick. The efficiency of this type of support for the assay of tritiated thymidine, ($i.e.$ 35—40%), is reported[84] to be significantly higher than that of paper discs (1—2%), or glass fibre discs (5—10%).

An ingenious method for the treatment of thin layers for assaying in liquid scintillation counting systems has been described.[85] The problem with many commercial methods of 'fixing' the thin layer support is the need to use water for the removal of the layer, with a consequent risk of loss of water-soluble material from the support. The spray mixture (Stripmix) suggested by these authors consists of 7 g cellulose acetate (BDH), 3 g diethylene glycol, 2 g camphor, 25 cm³ n-propanol, and 75 cm³ acetone. A pool of $ca.$ 20 cm³ is spread evenly with a glass rod over the plate, and the surface dried for 5 to 10 min. When a section is cut, it curls off the plate and can readily be taken up by a forceps.

The measurement of higher energy β-emitters such as ^{32}P, on solid supports, requires particular attention to be paid to the orientation of the disc in the liquid scintillation counting system[86] since the amount of liquid scintillation fluid between the sample and the glass wall of the vial becomes a critical parameter in determining the efficiency of counting.

A 'general' solution to the assaying of materials absorbed onto solid supports, so that reproducible and accurate data can be obtained, has been described.[87] This method is an in-vial procedure and involves using scintillant mixtures containing detergents. Several different solid supports were examined in this study, and consistent data were obtained only when the samples stayed completely within the solid support or were completely eluted. Difficulties arose when partial elution occurred.

A useful observation has been made[88] with regard to the assay of $Ba^{14}CO_3$ resulting from trapping $^{14}CO_2$ gas. This is normally assayed in suspension, with the usual self-absorption artefactual difficulties, but Larsen[88] has pointed out that 1—6 mg of the $Ba^{14}CO_3$ may be easily dissolved in 1 cm³ of 0.05M edta tetrasodium salt in-vial and by then adding 10 cm³ of toluene–Triton X100 (2:1) scintillator, a clear colourless counting mixture of a colloidal structure, can be obtained for counting. This mixture is not homogeneous and care should be exercised in the choice of

[84] M. J. Slobodic and R. W. Granlund, *Health Phys.*, 1974, **27**, 128.
[85] R. J. Redgewell, N. A. Turner, and R. L. Bieleski, *J. Chromatog.*, 1974, **88**, 25.
[86] E. Blasius and N. Sparmhake, *Internat. J. Appl. Radiation and Isotopes*, 1973, **24**, 301.
[87] R. M. McKenzie and B. K. Gholson, *Analyt. Biochem.*, 1973, **54**, 17.
[88] P. O. Larsen, *Internat. J. Appl. Radiation and Isotopes*, 1973, **24**, 612.

^{14}C-standard, as the completely non-polar [^{14}C]hexadecane may tend to overestimate efficiency due to its preferential incorporation into the scintillant-rich, hydrocarbon phase of the colloid.

13 Colloidal Scintillation Counting

One of the major advances in liquid scintillation counting technique over the past few years has been the introduction of colloidal scintillation counting systems. The principle of the technique is to bring into the closest juxtaposition, the primary scintillant solution and the sample solution itself, without phase unification in the strict sense. A detergent is used in this case, as a phase separator, retaining a colloidal structure, where the phases are submicroscopic or even organized into a liquid crystalline structure at the molecular level. Optimal counting of the system xylene:Triton X100:water appears to require the molecular composition 3:1:20, where the sample, in this case [^{3}H]H$_2$O is entirely in the aqueous phase.[89]

Clearly, the structure of the colloidal system will be highly dependent on the proportion of hydrocarbon (primary solvent), aqueous sample phase, and the nature and number of detergent molecules present. A useful representation of these type of systems by a triangular plotting often used in phase studies has been applied to the study of these systems.[90] The triangular plotting technique was applied to the study of liquid scintillation counting aspects of the system, first by Van der Laarse[91] for ground water and brackish solutions in connection with the study of oil movements, and by Fox[92] later for the study of a number of solutions employed in biochemistry.

The amphiphilic detergent found to be most useful in colloidal scintillation counting is Triton X100 (Rohm and Haas, Inc.). This is a non-ionic detergent containing mostly iso-octylphenoxy polyethoxyethanol. The chain comprises approximately ten ethoxy-units. Other closely related detergents, such as Triton X114[93] or Triton N101 have been recommended as superior to Triton X100, but a critical appraisal, using a phase-diagram technique, did not confirm these reports.[89]

The colloid scintillation counting system is of greatest value in the case where an aqueous solution of salts, proteins, or other solutes needs to be assayed and where addition of even a small volume of a blending agent would lead to precipitation of the solute and consequent gross quenching problems. The technique is especially valuable where an assay of many fractions needs to be made, *e.g.* samples obtained from density-gradient centrifugation and chromatography. The scintillant composition is improved by the use of a primary solute which is least susceptible to quenching such as butyl-PBD, [2-(4′-tert-butylphenyl)-5(4″-biphenylyl)-1,3,4-oxadiazole]. A detergent-based scintillant composition, based on this scintillant was recommended for the assay of caesium chloride and sucrose gradients by Dobrota and Hinton.[94] The recommended mixture consisted of 31.5 g butyl PBD in a mixture of 3 l toluene, 0.5 l methanol, and 1.5 l Triton X100. Up to 0.5 cm^3 of 2 M sucrose or of 60% w/w caesium chloride could be added to 10 cm^3 of scintillant provided that 1.5 cm^3

[89] B. W. Fox, unpublished results.
[90] P. A. Winsor, *Chem. and Ind.*, 1960, 632.
[91] J. D. van der Laarse, *Internat. J. Appl. Radiation and Isotopes*, 1967, **18**, 485.
[92] B. W. Fox, *Internat. J. Appl. Radiation and Isotopes*, 1968, **19**, 717.
[93] L. E. Anderson and W. O. McLure, *Analyt. Biochem.*, 1973, **51**, 173.
[94] M. Dobrota and R. H. Hinton, *Analyt. Biochem.*, 1973, **56**, 270.

of water was added at the same time. The final concentration of sucrose was below 0.5 mol l^{-1} and that of caesium chloride, less than 15% (w/w).

Owing to the multicomponent nature of the colloidal counting system many empirical formulations have been derived claiming higher efficiencies or 'improvements' in some other parameters, e.g. salt or macromolecule acceptability. A systematic approach towards determining the best conditions for the most sensitive assay for a given mixture of detergent, hydrocarbon solvent and aqueous solution has been outlined.[95,96] Subsequent empirical modifications have been reported, however, such as the use of NCS to increase stability of a toluene:Triton X100(2:1) system for trichloroacetic acid solutions,[97] and the incorporation of ethanol and ethylene glycol to a Triton X100:toluene mixture.[98] The latter system requires more detailed study using triangular plotting techniques, as well as a critical comparison of both polar and non-polar carbon-14 standards, before the claims of the use of external standard ratio can be justified. The use of tritium standards would exaggerate the deviations from homogeneity and would be a more critical means of comparison. The possibility of assaying at an even higher proportion of aqueous solution to scintillant has been described;[99] this system uses a mixture of AmPO [66% w/w lauryldimethylamine oxide and polyethyleneglycol-mono(p-'nonyl'-phenyl) ether (33%)], and a toluene-based scintillant consisting of 10 g PPO, 0.2 g dimethylPOPOP in 1 l toluene, in the ratio of 1 g of AmPO to 1.5 cm^3 of toluene scintillant. With this cocktail, it is possible to produce a mixture of 25% aqueous solution of 75% AmPO scintillant, and achieve high counting efficiency.

14 Correction Procedures, Artefacts, and Data Processing

All scintillation counting, in practice, contains a certain level of quenching. Although many methods of sample preparation seek to reduce the level of this artefact to a minimum, in most cases it is more convenient to measure the level of quenching in order to determine the level of interference present and hence to make an assessment of the absolute level of radioactivity present. Alternatively, the level of quenching may be used simply to equate samples from this point of view so that the level of counting efficiency of each sample is comparable.

Three methods which have been used to determine the level of quenching present will be considered. The first is internal standardization, involving the addition of a known amount of the sample labelled in a standardized manner, or failing this, a standard substance of similar polarity. The second is the sample-channels ratio method, which involves measuring the level of counts, derived from the isotope in the sample, seen in two selected channels. Any quenching effect, tends to shift the count rate down the energy spectrum and this appears to move counts from the upper channel to the lower channel. The ratio of the counts within these two channels therefore gives a good indication of the level of quenching. This procedure is very efficient provided the number of counts in the sample is sufficient to provide a statistically valid ratio when two counts are divided between the two selected

---

[95] B. W. Fox, *Internat. J. Appl. Radiation and Isotopes*, 1974, **25**, 209.
[96] B. W. Fox, in ref. 3, p. 202.
[97] P. N. P. Chow, *Analyt. Biochem.*, 1974, **60**, 322.
[98] U. Fricke, *Analyt. Biochem.*, 1975, **63**, 555.
[99] W. Lahmann and A. Hinzpete, *Internat. J. Appl. Radiation and Isotopes*, 1974, **25**, 515.

windows. To improve the statistics of the counting, the usual practice is to overlap the channels used in the ratio determination. The third method is an artificial source device, in which an external γ-emitter is mechanically introduced near the vial containing scintillant and sample and the induced Compton electrons are used to produce photons in much the same way as an internal β-emitter.

The counting instrument, therefore, can be designed to determine the quench shift from either this method or the sample-channels ratio method and to calculate the degree of quenching. The accuracy of the external standardization method over a wide range of sample and vial parameters appears to be more reliable for the higher energy γ-emitters such as ^{226}Ra rather than the lower energy ^{137}Cs.[100] The relationship between quench-correction techniques and the vial material has been the subject of a number of reports since the original observation of Rauschenbach and Simon[101] that, with polyethylene vials, the external standard ratio underwent a temporal variation when low-energy external standard sources (*e.g.* ^{137}Cs) were used. Such variations were not observed with the higher energy sources (^{226}Ra).

Within a single sample there is often both colour and chemical (impurity) quenching and in order to determine the contribution of each, a simple method employing concentric tubes and an internal standard consisting of an internal conversion electron emitter system, ^{113}Sn–^{113}In is used.[102] A non-quenching system is produced by using a toluene solution of diphenyloxazole in the inner tube and toluene alone in the outer. Total quenching is obtained by adding quencher to the inner tube and the colour quenching component is obtained by adding a colour quenching agent to the outer compartment alone. In the study[102] it was shown that the extent of colour quenching corresponds to the degrees of overlap of the quencher absorption spectrum and the scintillator emission spectrum. Impurity quenching, on the other hand, could not be related to any structural features over the 30 different quenchers examined, except that they exhibited a concentration dependence as well as some similarities between isomers.

Another study attempting to relate the structure of many carotenoids to the degree of colour quenching[103] produced two interesting results. Firstly, all the data from all the carotenoids studied fell on the same external standard ratio–percentage efficiency curve. Secondly, attempts to correlate the degree of colour quenching by individual carotenoids with their light-absorption characteristics and the scintillation emission spectrum of the scintillator, led to the unexpected and anomalous observation that the pigments which absorb maximally in the region of 350—450 nm are not necessarily the most powerful quenchers. The attenuation of photon pulses due to light absorption creates a geometry effect in colour quenching, a factor which has now been studied.[104]

A method reported[105] for the determination of a chemical quench curve consists of allowing a slow evaporation of carbon tetrachloride within a vial, while repeated counts are made on the scintillator situated below the evaporator mechanism. This method cannot, of course, be applied to colour quenching.

[100] F. Gogan and P. Gogan, *Analyt. Biochem.*, 1974, **60**, 363.
[101] P. Rauschenbach and H. Simon, *Z. analyt. Chem.*, 1971, **256**, 119.
[102] M. Takine and H. Ishikawa, *Nuclear Instr. and Methods*, 1974, **118**, 51.
[103] P. M. Bramley, B. H. Davies, and A. F. Rees, in ref. 3, pp. 76—85.
[104] F. E. L. ten Haaf, in ref. 3, pp. 41—43.
[105] M. A. Reunanan and E. J. Souri, in ref. 3, pp. 86—93.

Heterogeneity of the scintillator system can lead to artefacts if the external standard ratio is used to correct for quenching. Where heterogeneity is suspected the correction should be carefully checked with the sample-channels ratio, and if discrepancy is found, the latter method should be used for quench correction.

The problems associated with the assay of β-particles in heterogeneous systems have been the subject of a number of interesting studies. Considerable distortion of the energy spectrum of a β-emitter has been shown[106] to occur, when the sample is on solid supports. These distortions are also related to the positioning of the sample within the vial and arise from the fact that the ranges of the β-particles from certain positions, e.g. near the vial wall and on the bases of the vial, can exceed the depth of the scintillator solution available to them. This effect increases with the energy of the β-particle. This phenomenon can lead to variability in the normal counting channels and considerable difficulties, especially if double-isotope assays are attempted.

It is difficult to see how such a variable feature as sample absorption onto surfaces can be accounted for when quenching needs to be determined. However, the problems associated with wall absorption have been the subject of corrective procedures.[107] An appropriate method of overcoming the counting heterogeneity involved is to dilute the β-emitting particle with unlabelled material of the same chemical form as that of the β-emitter undergoing absorption, and hence to decrease considerably the specific activity of the material adhering to the wall surface. An alternative method[108] takes into consideration the two counting efficiencies E_{A-}, expected when absorption artefacts are excluded and E_{A+}, expected if the sample were fully absorbed on the surface. The activity expressed in d.p.m. would then be

$$= \frac{\text{c.p.m} \times 10^4}{E_{A-} \times E_{A+}}$$

The adsorption efficiency is determined from a calibration curve using an adsorption shift factor (A), which varies from a value of 0 in a completely non-adsorbed sample to 0.4 in a fully adsorbed sample and is defined by the equation

$$A = a \log x + b - \log y$$

where x is the external standard ratio, y is the sample-channels ratio and a and b are experimentally determined constants, which are determined from a series of quenched standards of non-adsorbed samples. The adsorption shift factor (A) is determined from the deviation (in terms of log y) from the non-adsorbed curve.

The structural aspects of a wide variety of organic compounds have also been studied by Wigfield and Srinavasan,[109] using their adsorption shift factor as a criterion. Of 44 radiochemicals examined, 11 showed adsorption effects. Amongst the latter, it appears that the presence of oxygen in the molecule is necessary for adsorption, and that dicarboxylic acids are particularly prone to adsorption, whereas monocarboxylic acids are not. In order for an adsorption-prone chemical

---

[106] F. G. Winder and G. R. Campbell, *Analyt. Biochem.*, 1974, **57**, 477.
[107] D. C. Wigfield, *Analyt. Biochem.*, 1974, **59**, 11.
[108] D. C. Wigfield and V. Srinavasan, *Internat. J. Appl. Radiation and Isotopes*, 1973, **24**, 613.
[109] D. C. Wigfield and V. Srinavasan, *Internat. J. Appl. Radiation and Isotopes*, 1974, **25**, 473.

to be adsorbed, it must have at least 0.16 mCi mM^{-1} specific activity (this value was determined for glass vials). This specific activity corresponds to *ca.* 6 × 10^{14} molecules cm^{-2} of glass surface.

An interesting observation has been reported[110] involving the degradation of highly polymeric tritiated methyl methacrylate in butyl PBD scintillant. The degradation was not due to the level of tritium in the polymer, but it appeared more probable that it was due to the fluorescence emission of the scintillant itself.

Where homogeneity of the sample preparation is assured, the external standard ratio is an ideal system for correction of quenching, especially by simple computer methods. There have been many reports of computer programs devised to undertake this correction and determine the d.p.m. Several have ingenious short cuts to overcome the limited memory facilities of many desk computers. The computation of the typical external standard ratio–efficiency curve is usually based on fitting a curve to a second or third order quadratic. In practice, such equations have not accurately described the actual curve obtained. A better method appears to be to define the curve as a series of short lines and a simple program based on this principle for use with the Olivetti 101 calculator has been described:[111] programs are given to determine the slopes of the lines and their intercepts, as well as the d.p.m. Double labelled, variably quenched samples provide a suitable subject for analysis with computer assistance and a Fortran IV program SCINT[112] and an ALGOL program[113] have been designed with a variety of applications.

The assay of triple-labelled material, *e.g.* ^{3}H, ^{14}C and ^{36}Cl is often required in plant physiology and a suitable off-line data analysis system using the Olivetti P203/06 desk computer with a tape input unit has been described.[114] A flexible Fortran IV program, with optional features for the user's specific requirements has also been described[115] for the processing of liquid scintillation counting data.

Although many studies have been conducted into possible sources of error in techniques or instrumentation, the difficult area of operator error has been little studied, although some indication that these errors may be exceptionally great was obtained by Scales.[116] Large within-operator errors were obtained, even with commercially standardized combustion methods.

15 Chemiluminescence and Phosphorescence

In scintillation counting, the phenomenon of chemiluminescence has been a considerable embarrassment to many biomedical assay applications. In particular, this artefact often follows attempts to solubilize liver and blood-containing samples of tissues with quaternary ammonium solubilizers, especially following freeze-drying procedures of the tissues, or their homogenates.

The spurious counts obtained in these type of experiments have been investigated by Laine–Böszörmenyi and Fallot[117] who have shown them to be primarily associa-

[110] J. F. Norris and F. W. Peaker, *Internat. J. Appl. Radiation and Isotopes*, 1974, **25**, 143.
[111] H. J. Hope, *Analyt. Biochem.*, 1973, **53**, 295.
[112] B. L. Boeckz, D. J. Protti, and K. Dakshinamurti, *Analyt. Biochem.*, 1973, **53**, 491.
[113] J. Sliwowaki and G. Wozniakowska, *Roczniki Chem.*, 1973, **47**, 2151.
[114] H. Veen, *Internat. J. Appl. Radiation and Isotopes*, 1974, **25**, 355.
[115] D. E. Bowyer and J. D. Pearson, in ref. 3, pp. 94—106.
[116] B. Scales, in ref. 3, p. 211.
[117] M. Laine-Böszörmenyi and P. Fallot, *Internat. J. Appl. Radiation and Isotopes*, 1974, **25**, 241.

ted with the presence of auto-oxidizable lipids which increase with storage in cooled temperature conditions. The lipid-free fraction of the tissues were shown to produce considerably less chemiluminescence than the lipid-containing extracts. The slight chemiluminescence associated with the lipid-free fraction had different kinetics from that associated with the lipid fraction, and is of unknown origin. The authors suggest that such tissues are best extracted with methanol:chloroform (1:2) mixtures before assay except where the label occurs in the lipid fraction.

16 Biochemical Applications

The technique of liquid scintillation counting has found its most important applications in the fields of biochemistry and bio-medicine.

The assay of ^{55}Fe and ^{59}Fe in blood has been repeatedly examined and reported over the last decade. Attempts to decrease the labour involved in sample preparation in which the iron is concentrated and decolourized have been frequently reported. A simple procedure[118] consisting of a decolourization step with sodium hypochlorite followed by colloidal scintillation counting (Instagel–Packard) at a reduced temperature has enabled whole blood to be assayed rapidly and accurately for the two iron isotopes.

The double-isotope derivative technique, whereby one isotope is used to label an internal standard and another to label the reagent, is a well established method in both biochemistry and physiological chemistry, but it has not been widely exploited in pharmacological applications. Riess[119] has outlined the method, using the assay of maprotiline as an example. The principle of the procedure is that an aliquot of the material to be analysed (*e.g.*, blood, plasma, urine, tissue homogenate, *etc.*) is homogeneously mixed with ^{14}C-labelled drug, in this case, maprotiline. By a suitable isolation procedure, (extraction, chromatography, *etc.*) the ^{14}C label is extracted from the bulk and a derivative prepared with a tritium-labelled reagent, *e.g.* in this case, tritium-labelled acetic anhydride. The tritium derivative is purified by extraction or thin layer chromatography. Each sample in the final purification step is analysed for both tritium- and ^{14}C-activity using the liquid scintillation counting procedure. Standards allow the calculation of ^{14}C yields and the recovered tritium activity can be corrected to a theoretical 100% yield. The resultant tritium activity value can then be converted into concentration units by comparison with 100% tritium values obtained from a series of test samples containing known definite quantities of drug, treated under the same conditions.

The assay of radioactive metabolites in faeces has presented a number of problems. Of importance is the volatility of the metabolite. If it is non-volatile, then a sample obtained from dried faeces can be subject to either solubilization or extraction procedures. In either case, the radioassay is often difficult owing to excessive colour quenching.

The assay of [^{3}H]- or [^{14}C]cholesterol in faeces is an example of a situation where the faeces can be initially extracted with petroleum ether; an aliquot of the extract is then evaporated to dryness within the scintillation vial using gentle warming and a stream of nitrogen gas. Decolourization is then effected by adding 10 drops of a

---

[118] A. Wagner, *Internat. J. Appl. Radiation and Isotopes*, 1973, **24**, 548.
[119] W. Riess, *Analyt. Chim. Acta*, 1974, **68**, 363.

5.25% solution of sodium hypochlorite[120] to the vial and the mixture is swirled so that all the sample and glass surfaces are covered. Decolourization occurs within 24 h but the best results are obtained after 48 h. Ten drops of 0.1 M sodium nitrite solution are then added and the mixture is again swirled, the reaction being slightly exothermic. The product is then blended into a water-accepting scintillant mixture such as Aquasol or Bray's solution. The assay of [^{14}C]cholesterol in plasma or tissue homogenate, where colour quenching does not present a problem, is more simply conducted by adding up to 500 μl of the sample directly to a scintillation vial and partitioning the lipids directly into 15 cm^3 of a toluene-based scintillation mixture containing 37.5% ethylene glycol monomethyl ether. By adding 1 cm^3 of water, two phases form, the upper toluene scintillant phase containing all the cholesterol.[121] Bleaching is unnecessary in this method.

An interesting application of the isotope-dilution technique has been described for the assay of specific amino-acids in a mixture.[122] The method involves adding an exact amount of labelled amino-acid to the non-radioactive amino-acid mixture to be analysed, and then adding a limiting quantity of t-RNA in the presence of its synthetase and ATP and Mg^{2+}. The enzyme reaction is stopped by addition of trichloroacetic acid and the resulting macromolecular precipitate is filtered onto glass fibre discs, washed, dried, and assayed.

A sensitive assay for the proteolytic activity of certain enzymes such as trypsin and papain has been described.[123] The synthesis of the specific substrate, benzyl-DL-arginine [^{3}H]anilide hydrochloride is given. The enzyme reaction is carried out in the scintillation vial in an aqueous layer, the toluene-soluble product entering the toluene-scintillant phase above. Repeated 1 min counts on the mixture enable the time-course of the enzyme reaction to be obtained and the kinetic parameters of the reaction to be calculated. Diffusion of the product into the toluene layer is so fast that shaking is unnecessary.

A rapid and simple assay for the enzyme, thymine-7-hydroxylase has also been reported.[124] The method consists of the conversion of tritiated thymine into 5-hydroxymethyluracil by the enzyme. The loss of tritium atoms as tritiated water is proportional to the level of the 5-hydroxy-derivative formed. The procedure can be increased in sensitivity by the addition of catalase and bovine serum albumin to the reaction mixture. A similar assay[125] of microsomal aryl hydroxylase depends on the liberation or tritiated water from [$U - {}^3$H]benz(α)pyrene. As little as 3 g of rat-liver microsomes is required for the assay and a celite column in a pasteur pipette is used to absorb everything but the liberated water.

The direct determination of the specific activities of [^{14}C]proline and [^{14}C]-hydroxyproline in collagenous as well as non-collagenous proteins has been described.[126] The method consists of treating a 2 cm^3 sample with 1 cm^3 of chloramine T for 20 min at room temperature, followed by addition of 0.5 cm^3 of 2 M sodium thiosulphate to stop the reaction. After thorough Vortex mixing, 1.0 cm^3 of 1 M

[120] L. M. Davidson and D. Kritchevsky, *Analyt. Biochem.*, 1975, **66**, 287.
[121] J. Vikari, *Analyt. Biochem.*, 1975, **63**, 566.
[122] K. Beaucamp and H. E. Walter, *F.E.B.S. Letters*, 1973, **38**, 37.
[123] S. Roffmann and W. Troll, *Analyt. Biochem.*, 1974, **61**, 1.
[124] T. Z. Liu, C. H. Wong, and S. B. Shohet, *Analyt. Biochem.*, 1974, **62**, 408.
[125] T. Hayakawa and S. Udenfriend, *Analyt. Biochem.*, 1973, **51**, 501.
[126] M. Rojkind and E. Gonzalez, *Analyt. Biochem.*, 1974, **57**, 1.

NaOH is added, the mixture shaken and saturated with NaCl (2.0 g). Toluene (6 cm^3) is added and the oxidation product formed containing the ^{14}C is extracted within 30 s. An aliquot (4.0 cm^3) of this toluene layer is transferred directly to a scintillation vial for assay.

In many types of experiments involving the administration of tritiated compounds to animals, the possibility of specific uptake of the tritium atom itself has been assumed to be negligible. An experiment has been conducted[127] to demonstrate that this is in fact so. Tritiated water was fed to rabbits for three generations at the rate of 1 mCi cm^{-3}. In addition, the food, namely Alfalfa, was grown by hydroponics with the same specific activity of tritiated water, so that the entire intake of water from any source was tritiated to the same extent. It was found that no selective uptake of the tritium occurred and that the specific activity of the hydrogen component of the organic material from all tissues was the same as the tritiated water consumed.

17 Radioimmunoassay and Related Techniques

Radioimmunoassay (RIA), radioreceptor assays, protein binding assays, and other saturation assays have provided the basis of the most significant advances in biomedicine during the period of this review. Useful reviews have been prepared[128—130] as well as a report of the International Atomic Energy Agency Panel on Standardization of Radioimmunoassay procedures,[131] and it is clear that many of the earlier applications in this field have employed γ-emitters more suitable for γ-counting instruments. However, apart from the numerous RIA procedures which use ^{3}H- and ^{14}C-labelled material necessitating liquid scintillation procedures, many of the gamma emitting isotopes can now be measured by modified liquid scintillation counting procedures. Many of these have also been measured with a high efficiency.[132,133]

Some examples of specific ligand assays have been referred to above in connection with enzyme assays. In the RIA procedure, the antigen being measured (usually an aliquot of plasma) is mixed with a known amount of labelled antigen and a fixed amount of antibody. The labelled and unlabelled antigens compete for the limited antibody sites (ideally sufficient to bind approximately 50% of the labelled antigen). On equilibrium, the labelled and unlabelled antigens will be partly bound and partly unbound to antibody. The unbound components are usually adsorbed onto charcoal or resin and the level of bound radioactivity in the supernatent fluid is measured by a counting technique. The antigen can be a protein or it can be a specific small molecule hapten, bound to a labelled protein, and it is within this latter application that considerable advances have occurred.

Commercial kits are now available for many applications of this technique and three of these for serum digoxin using a tritium tracer have been compared.[134]

[127] A. A. Moghissi, R. E. Stanley, J. C. McFarlane, E. W. Bretthauer, R. G. Patzer, and S. R. Lloyd., *Radiation Res.*, 1974, **59**, 63.
[128] Radioimmunoassay and Saturation Analysis, ed. P. H. Sonksen, *Brit. Med. Bull.*, 1974, **30**, (1).
[129] E. Z. Helman and P. Ting, *Clin. Chem.*, 1973, **19**, 191.
[130] J. E. McGuigan, *Mayo Clinic Proc.*, 1973, **48**, 637.
[131] I.A.E.A. Panel Report, *Internat. J. Appl. Radiation and Isotopes*, 1974, **25**, 145.
[132] D. L. Horrocks, in ref. 3, pp. 28—33.
[133] H. B. Herscowitz and T. W. McKillip, *J. Immunol. Methods*, 1974, **4**, 253.
[134] N. P. Kubasik, S. Schauceil, and H. E. Sins, *Clin. Biochem.*, 1974, **7**, 206.

Of the many methods for the assay of protein and other macromolecular substances, several have already been reviewed, such as posterior pituitary peptides[135] and carcinoembryonic antigen.[136] RIA methods for the renin–angiotensin system,[137—140] angiotensin in human blood,[141] human chorionic somatomammotropin,[142] thyrotropin-releasing hormone,[143] and human serum albumin in peritoneal dialysate[144] have been described. In addition the level of gonadotrophin in salmon has also been estimated by this procedure.[145]

RIA has probably made its greatest impact in endocrinology and a review of the use of the technique in understanding human adrenal physiology has been prepared.[146] RIA methods have been used for plasma[147—153,139] and urinary[154] aldosterone, plasma testosterone,[155—159] for testosterone and dihydrotestosterone simultaneously,[160] plasma oestriol,[161] oestradiol,[162,163] progesterone[164] as well as for endocrinologically active agents such as 17-ethynylestradiol and mestranol[165] and Provera (medroxyprogesterone acetate).[166]

RIA methods have been used for drugs such as Pentazocine,[167] Dexametha-

[135] A. G. Robinson and A. G. Frantz, *Metabolism*, 1973, **22**, 1047.
[136] U. Stevens and D. J. R. Laurence, *Annales d'Immunol.*, 1973, **124c**, 615.
[137] S. Ulick, *Metabolism*, 1973, **22**, 1027.
[138] M. A. Waite, M. Tree, and E. A. McDermott, *J. Endocrinol.*, 1973, **57**, 329.
[139] P. Corvol, J. Menard, and X. Bertagna, *Ann. Endocrinol.*, 1973, **34**, 57.
[140] J. M. McDonald and G. A. Fischer, *Amer. J. Clin. Pathol.*, 1973, **59**, 858.
[141] M. A. Waite, *Clin. Sci. Mod. Med.*, 1973, **45**, 51.
[142] F. Cocola, A. R. Genazzani, and P. Neri, *J. Nuclear Biol. Med.*, 1973, **17**, 14.
[143] S. L. Jeffcoate, H. M. Fraser, A. Gunn, and N. White, *J. Endocrinol.*, 1973, **59**, 191.
[144] R. Bianchi, G. Mariani, S. Masini, F. Matera, and G. C. Zucchelli, *J. Nuclear Biol. Med.*, 1974, **18**, 112.
[145] L. W. Crun, P. K. Meyer, and E. M. Donaldson, *Gen. and Comp. Endocrinol.*, 1973, **21**, 69.
[146] C. D. West and F. H. Tyler, *Metabolism*, 1973, **22**, 995.
[147] B. T. Martin and C. A. Nugent, *Steroids*, 1973, **21**, 169.
[148] N. Varsano-Aharon and S. Ulick, *J. Clin. Endocrinol. and Metabolism*, 1973, **37**, 372.
[149] C. A. Bizollon, J. F. Riviero, and B. Claustrat, *Ann. Endocrinol.*, 1974, **35**, 725.
[150] J. Y. Daniel, H. Mion, and D. Soulas, *Clin. Chim. Acta*, 1974, **55**, 235.
[151] R. W. Farmer, D. H. Brown, P. Y. Howard, and L. F. Fabre, jun., *J. Clin. Endocrinol. and Metabolism*, 1973, **36**, 461.
[152] C. Gomez-Sanchez, D. C. Kem, and N. M. Kaplan, *J. Clin. Endocrinol and Metabolism*, 1973, **36**, 795.
[153] P. Giannotti, M. Mannelli, G. Feorelli, and M. Serio, *J. Nuclear Biol. and Med.*, 1974, **18**, 104.
[154] J. Stuart, P. M. Keane, G. Viol, and R. N. Gupta, *Clin. Biochem.*, 1973, **6**, 285.
[155] M. Luise, F. Franchi, G. F. Menchini, D. Barletta, C. Farsorra, F. Ciardella, and G. Gagliardi, *Steroid and Lipid Res.*, 1973, **4**, 213.
[156] H. L. Verjans, B. A. Cooke, F. H. deJong, C. M. M. deJong, and H. J. van der Molen, *J. Steroid Biochem.*, 1973, **4**, 665.
[157] A. Castro, H. Shih, and A. Chung, *Experientia*, 1973, **29**, 1447.
[158] A. Castro, H. H. Shih, and A. Chung, *Steroids*, 1974, **23**, 625.
[159] J. P. P. Tyler, J. G. Hennam, J. R. Newton, and W. P. Collins, *Steroids*, 1973, **22**, 871.
[160] J. M. Barberia and I. H. Thorneycroft, *Steroids*, 1974, **23**, 157.
[161] E. Youssefnejadian and I. F. Sommerville, *J. Steroid Biochem.*, 1973, **4**, 654.
[162] L. Verdonck and A. Vermeulen, *J. Steroid Biochem.*, 1974, **5**, 471.
[163] B. G. England, G. D. Niswender, and A. R. Midgley, jun., *J. Clin. Endocrinol. and Metabolism*, 1974, **38**, 42.
[164] B. Hoffmann, H. J. Kyrein, and M. L. Ender, *Hormone Res.*, 1973, **4**, 302.
[165] P. N. Rao, A. de la Pena, and J. W. Goldzieher, *Steroids*, 1974, **24**, 803.
[166] M. E. Royer, K. Ko, J. A. Campbell, H. C. Murray, J. G. Evans, and D. G. Kaiser, *Steroids*, 1974, **23**, 713.
[167] T. A. Williams and K. A. Pittman, *Res. Comm. Chem. Pathol. Pharm.*, 1974, **7**, 119.

zone,[168] Adriamycin and Daunomycin,[169] phenobarbitone,[170] amphetamines,[171] prednisolone;[172] a review of the methods available for the digitalis glycosides[173] has also appeared.

An interesting application of the RIA technique for the assay of specific nucleic acid bases, *viz*. N^2-dimethyl-guanosine, 7-methylguanosine, and pseudouridine has also been reported.[174] In addition, the assay of cyclic nucleotides and the associated cyclase activity has been described.[175] Serum thymidine has been assayed by RIA[176] in a method based on the competition of thymidine and [^{125}I]iododeoxyuridine for thymidine antibody sites. The level of antibody-bound [^{125}I]IUdR was measured in ammonium sulphate precipitates. An interesting observation made by these workers is that the level of serum thymidine increases in rats, following X-irradiation.

The use of competitive binding of hormones to proteins, not employing immunological methods, but relying on specific protein ligands, has been reviewed.[177] The development of such assay techniques for the analysis of Vitamin B12, folic acid, and its analogues has been described. The method consists of diluting the ligand (*e.g.* enzyme) until competitive binding between labelled and unlabelled vitamin is achieved.

The potentially wide application of RIA to the assay of vitamins may be ascertained from a review of the subject.[177] Methotrexate, a drug which binds specifically to the enzyme dihydrofolate reductase can also be assayed by RIA. The measurement of serum-folate and folic acid binding proteins using tritiated folic acid in this type of assay has been described.[178]

When the compound to be measured is totally bound to specific antibodies, the method is usually referred to as an immunoradiometric assay. This type of assay has been reviewed by Woodhead *et al.*[179] An example of this technique is the measurement of serum ferritin.[180]

A three-parameter mathematical model has been devised[181] for the fitting of experimental data from competitive protein binding assays, in particular, from those methods used in the estimation of steroid hormones.

18 Archaeological and Hydrological Applications

The use of liquid scintillation counting in archaeology and hydrology arises from considerably improved techniques of sample preparation, in particular, the concentration of carbon-14 and tritium into solvent materials such as benzene which

[168] M. Hichens and A. F. Hoyans, *Clin. Chem.*, 1974, **20**, 266.
[169] H. V. Van Vunakis, J. J. Langone, L. J. Riceberg, and L. Levine, *Cancer Res.*, 1974, **34**, 2546.
[170] A. Chung, S. Y. Kimm, L. T. Cheng, and A. Castro, *Experientia*, 1973, **29**, 820.
[171] L. T. Cheng, S. Y. Kim, A. Chung, and A. Castro, *F.E.B.S. Letters*, 1973, **36**, 339.
[172] W. A. Colburn and R. H. Buller, *Steroid*, 1973, **21**, 833.
[173] T. W. Smith and E. Haber, *Pharmacol. Rev.*, 1973, **25**, 219.
[174] L. Levine and H. Gjika, *Arch. Biochem. Biophys.*, 1974, **164**, 583.
[175] A. L. Steiner, *Pharmacol. Rev.*, 1973, **25**, 309.
[176] W. L. Hughes, M. Christine, and D. Stollar, *Analyt. Biochem.*, 1973, **55**, 468.
[177] S. P. Rothenberg, *Metabolism*, 1973, **22**, 1075.
[178] S. Waxman and C. Shreiber, *Blood*, 1973, **42**, 281.
[179] J. W. Woodhead, G. M. Addison, and C. N. Hales, *Brit. Med. Bull.*, 1974, **30**, 44.
[180] L. E. M. Miles, D. A. Lipshitz, C. P. Bieber, and J. D. Cook, *Analyt. Biochem.*, 1974, **61**, 209.
[181] A. Arrigucci, E. Calabresi, G. Fiorelli, G. Forti, M. Pazzagli, and M. Serio, *J. Nuclear Biol. Med.*, 1973, **17**, 124.

Sample Preparation Procedures for Liquid Scintillation Counting 131

can be used as high efficiency electron-trapping solvents in liquid scintillation counting. An important transition in radiocarbon dating technology has been the change from gas proportional assay to liquid scintillation counting.

The Mk II version of the bomb combustion system for radiocarbon analysis[48] incorporates a silica-insulated electrode and two heat shields in a water-cooled steel cylindrical vessel with a removable 100 atm bursting disc (0.005 cm thick) and gas inlet and outlet systems. The bomb body is 38 cm long with an overall diameter of 14 cm with 1.3 cm thick walls and a capacity of 3.9 l. Prior to combustion of approximately 12 g of carbon, 100 cm^3 of degassed water is injected to remove any strongly acidic gases.

Natural tritium is widely used as a means of estimating the movement of water in both geology and meteorology. The technique depends largely on the ability to which the level of tritium present in a water sample can be enriched to measurable levels by physical means. The radioassay of environmental tritium has been much improved by initially converting[182] the water into acetic acid by reaction with acetic anhydride. The [^{3}H]acid is then used to prepare tritiated acetylene by reaction with calcium carbide. The acetylene polymerization to benzene is then conducted in the same way as for radiocarbon. Levels less than 1 TU have been claimed to be detected by this procedure.

19 Miscellaneous Applications

Liquid scintillation counting has been used in many fields other than the biomedical and inorganic for which its greatest use in the past has been found. Its application to radioimmunoassay and related techniques has shown that once a successful application in a field has been demonstrated, its use can expand quickly to fill the technological gap.

Another such application is in the monitoring of the environment for contamination by α- and β-emitters, whether in an experimental laboratory or in effluent from nuclear reactor establishments. In both cases such monitoring has become obligatory. The measurement of levels of fall-out demands increasingly more sensitive assays. Such assays need to be especially sensitive in the field of natural isotope monitoring. The monitoring of air filters for α- and β-emitters has been described,[183,184] filters up to 320 cm^2 were assayed with a high figure of merit. The use of liquid scintillation counting methods in the measurement of α-emitting isotopes have already been referred to and the measurement of low levels of uranium[185] as well as gross α-activity in water and urine[186] are examples of the use of these methods. In addition a more detailed analysis of the isotopes has been made by exploiting the principles of Pulse Shape Discrimination,[187–189] thus extending considerably the potential of the system. With the application of specialized solvent-

---

[182] M. Wolf, *Internat. J. Appl. Radiation and Isotopes*, 1973, **24**, 299.
[183] K. Buchtela and M. Tschurlovits, *Health Phys.*, 1974, **27**, 131.
[184] K. Buchtela, *Nuclear Instr. and Methods*, 1974, **120**, 203.
[185] D. L. Horrocks, *Nuclear Instr. and Methods*, 1974, **117**, 589.
[186] M. B. Hafez and F. I. A. Saied, *Internat. J. Appl. Radiation and Isotopes*, 1973, **24**, 241.
[187] J. H. Thorngate, W. J. McDowell, and D. J. Christian, *Health Phys.*, 1974, **27**, 123.
[188] J. W. McKlvean, *Health Phys.*, 1974, **27**, 626.
[189] K. Buchtela, M. Tschurlovits, and E. Unfried, *Internat. J. Appl. Radiation and Isotopes*, 1974, **25**, 551.

extraction processes to these techniques[190] a particularly useful tool for a more detailed analysis of complex mixtures of isotopes has become available.

Liquid scintillation counting has also been applied in somewhat unexpected fields such as in the investigation[191] of the distribution of oils in both petrol and diesel engines. The generally tritiated oil is diluted into toluene and an aliquot is measured in a 0.6% solution of BBOT in toluene. The investigation of leaks in gas-containing equipment has also been studied by the technique. In this case, ^{85}Kr has been used. Its γ-emission allows it to be measured by conventional sodium iodide crystal scintillation counting, but since this emission is only 0.41% of the total emission, considerably greater sensitivity has been achieved[192] by methods which can also measure Cerenkov and bremmstrahlung radiation.

An unusual application in pharmacology is the investigation of the relationship between the psychosomatic action of a drug and its physico-chemical partitioning properties between oil–water interfaces.[193]

In marine biology, the 'assimilation' and 'mineralization' of sea water samples have been measured, by investigating the extent by which a sample of labelled glucose is metabolized to $^{14}CO_2$. Conventional trapping in Hyamine[194] and assay in toluene-based scintillants were used.

In the petroleum industry, the use of liquid scintillation counting in the assay of the products of alkylation and cracking reactions, initially separated by gas chromatography, has been described.[195] Gas adsorption onto a platinum surface, important in such hydrocarbon work, has also considerable theoretical interest in electrochemistry. The reaction has been subject to liquid scintillation counting applications[196] and useful fundamental information in this field has been obtained.

Clearly there is a potentially wide field of application for this technique and it is likely that many further uses will be described as increased sensitivity becomes a major criterion in the investigation. The use of isotopes by an increasing number of countries also makes it imperative that reliable and sensitive measuring techniques become available to monitor and help control the levels of these isotopes in the biosphere.

[190] J. W. McDowell, *Health Phys.*, 1974, **27**, 626.
[191] R. Evans, *Internat. J. Appl. Radiation and Isotopes*, 1973, **24**, 19.
[192] E. Brumix, *Internat. J. Appl. Radiation and Isotopes*, 1973, **24**, 359.
[193] H. Kumizuka and L. G. Abood, *J. Pharm. Sci.*, 1973, **62**, 740.
[194] A. M. Herbland and J. F. Bois, *Marine Biol.*, 1974, **24**, 203.
[195] L. H. Handler, and E. W. Smith, *Internat. J. Appl. Radiation and Isotopes*, 1974, **25**, 521.
[196] A. Czerwinski, J. Sobkowski, and A. Wiechowski, *Internat. J. Appl. Radiation and Isotopes*, 1974, **25**, 295.

Author Index

Aaij, C., 96
Abascal-M., R., 38, 42, 44, 64
Abood, L. G., 132
Abu-Samra, A., 71
Acerbi, E., 92, 106
Adams, A. B., 68
Addison, G. M., 130
Agudo, E. G., 19, 26
Ahmad, I., 26
Akalin, O., 26
Akerman, K., 9
Akiha, F., 76
Akindinov, V. A., 27
Akopov, V. S., 1
Aladjem, A., 20
Alaerts, L., 26, 29
Al-Attiya, M. J., 25
Albee, A., 69
Alexandrescu, P., 44
Alfassi, Z. B., 75, 106, 107
Allaniyazov, M., 26
Allen, L. S., 23
Allen, R. O., 69
Allen, T., 13
Al-Sadir, J., 84
Al-Shahristani, H., 25
Alvarez, J., 102, 105
Alyavdin, S. V., 22
Ambrosino, G., 35
Amiel, S., 20
Anderson, L. E., 121
Andrae, W., 16
Andreev, V. K., 31
Anma, M., 31
Anpilogov, A. P., 31
Ansari, A., 82, 95
Arino, H., 101, 103, 107
Armstrong, F. E., 13
Arnfred, T., 115
Arnold, D. M., 31
Arriaga, C., 105
Arrigucci, A., 130
Artzy, L. M., 62
Asaro, F., 37, 52, 61, 62, 69
Ashcroft, J., 110
Aspin, N., 105
Aspinall, A., 39, 65
Astachowicz, J., 19
Aston, M. A. J., 4
Atkins, F., 107
Atkins, H., 82, 95, 96
Aydogdu, M. K., 32
Azaryan, A. A., 17

Babinet, D. D., 85
Baesskaya, G. M., 26

Bakels, C. C., 70
Balgna, I. P., 26
Balint, T. D., 17
Balla, B., 13
Bal'vas, Yu. P., 30
Bantcugil, E., 26
Banterla, G., 63
Barberia, J. M., 129
Barenbaum, A. A., 31
Barletta, D., 129
Barnes, W. E., 107
Barrandon, J. N., 68
Bartlett, W. G., 16
Basargin, N. N., 26
Basin, Ya. N., 30
Battaglia, D. J., 105
Baur, J. R., 109
Bazaniak, Z., 7
Bearden, A. J., 104
Beaucamp, K., 127
Beaver, J. E., 94
Beck, C. W., 68
Beck, R. N., 84
Beckage, T., 15
Beckstein, H., 1, 15
Beenboer, J. Th., 86
Beerling-van der Molen, H. D., 81, 84
Beihn, R. M., 86
Beil, R. G., 31
Bekerman, C., 84
Belenki, R. D., 31
Bellido, A. V., 26
Belyaev, Yu. B., 8
Belyakov, M. A., 27
Belyi, V. V., 30
Benaben, P., 68
Benedek, S., 23
Bennett, R. B., 37
Benson, P. H., 64
Bent, D. H., 60
Berchard, M., 27
Berger, H., 32
Berkutova, I. D., 25
Bermann, L. I., 30
Bertagna, X., 129
Beskin, L. I., 24
Beswick, C. K., 17
Beutner, H. P., 1
Beyer, C., 2
Bhattacharyya, S. N., 26
Biala, N., 9
Bianchi, R., 129
Bibby, D. M., 27
Biciolla, L., 26
Bieber, A. M., jun., 37, 44, 49, 62

Bieber, C. P., 130
Bieber, L. L., 117
Bieleski, R. L., 120
Bievelez, P., 103
Billinghurst, M. W., 103
Birattari, C., 92, 106
Bird, J. B., 71
Birdsall, R. L., 100
Birks, J. B., 108
Bishop, R. L., 59, 64
Bittel, R., 27
Bizollon, C. A., 129
Bland, W. H., 105
Blair, R. J., 105
Blasius, E., 120
Block, C., 25
Blomquist, S., 7
Blott, S. L., 25
Blue, J. W., 92, 107
Blumenauer, G. R., 32
Bluyssen, H., 67
Blyumentsev, A. M., 31
Bobowski, D. L., 113
Boeckx, B. L., 125
Boessma, J., 107
Bois, J. F., 132
Bonnevie-Svendson, M., 19
Borzyak, I. P., 30
Boswell, C. R., 22
Boutaine, J. L., 2, 32
Bovey, R. W., 109
Bowman, H. R., 52, 69
Bowman, R., 45
Bowyer, D. E., 125
Bowles, K. J., 32
Boyce, I. S., 13
Boyd, R. E., 103
Bradley-More, P. R., 82, 107
Brafman, M., 9, 29
Braler, H. A., 25
Bramley, P. M., 123
Bramwell, S. E., 15
Bransome, E. D., jun., 111
Braun, G., 91
Braun, M., 25, 29
Braunstein, P., 107
Bray, G. A., 111
Brem, A. A., 22, 31
Bretthauer, E. W., 128
Brezeanu, G., 26
Brinkman, G. A., 76, 86
Bristow, J. D., 16
Britt, A. R., 117
Brooks, D. W., 37, 44
Brooks, H. L., 84
Brown, D. H., 129
Brown, D. J., 13

Author Index

Brown, L. C., 98, 107
Brown, M., 2
Brunhart, G., 107
Bruninx, E., 27, 132
Buchtela, K., 131
Buckingham, P. D., 77, 82, 85
Buddemeyer, E. U., 109
Buller, R. H., 103
Bullock, R. M., 8
Bumalova, A., 19
Burianova, M., 26
Burleigh, R., 115
Burton, R. A., 104
Bush, P., 70
Busick, D. D., 77
Butler, F., 111
Butler, J. W., 25
Bychvarov, N., 20

Calabresi, E., 130
Caldwell, R. L., 25
Caley, E. R., 33
Calhoun, F. L., 12
Calkins, R. C., 26
Callahan, A. P., 98, 107
Callahan, R. J., 105
Cameron, J. F., 13, 16, 19, 20, 23
Campbell, D. G., 26
Campbell, G. R., 124
Campbell, J. A., 129
Campbell, K. C., 25
Cann, J. R., 66
Carmichael, E., 68
Caro, R. A., 101
Carr-Brion, K. G., 15, 17, 19
Carré, C., 46
Carter, G. W., 116
Castagnet, A. C., 8
Castiglioni, M., 92, 106
Castranovo, F. P., 105
Castro, A., 129, 130
Catling, H. W., 34
Cauwe, F., 92
Cecal, A., 20
Cesarco, R., 19
Chamberlain, A. C., 106
Chan, L.-H., 41
Chan, P. K. H., 86
Chandra, R., 107
Chang, C. C., 85
Chao, S. E., 27
Chapin, F. S., 117
Chara, T., 8
Charlier, R., 103
Charlton, J. S., 1, 23
Chase, W. T., 41
Chauser, B. M., 94
Chen, L., 52
Chen, P. Y., 64
Chen, Y. M., 23
Cheng, L. T., 130
Chiba, M., 26
Chisholm, G. D., 94
Cho, B. Y., 16
Chow, P. N. P., 122
Christian, D. J., 131
Christine, M., 130
Christman, D. R., 73, 78, 82
Chruisciel, E., 24, 31
Chung, A., 129, 130
Ciardella, F., 129
Ciscato, V. A., 101

Clark, J. C., 77, 82, 85, 87
Clark, J. W. G., 20
Clark, L. P., 84, 85
Claustrat, B., 129
Clayton, C. G., 4, 17, 19, 20
Cleland, J. M., 115
Clough, W. S., 106
Cobean, R. H., 66
Coco, M., 102
Cocola, F., 129
Coe, M. D., 66, 68
Cogneau, M., 92
Cohen, I. M., 26
Cohen, M. B., 84, 85
Colburn, W. A., 130
Cole, C. M., 107
Collins, W. P., 129
Colombetti, L. G., 103, 107
Comar, D., 79, 81, 82, 107
Conforto, L., 70
Conlon, T. W., 10
Connell, G. M., 109
Constant, R., 103
Contafalska, M., 15
Conway, M., 71
Cook, J. D., 130
Cooke, B. A., 129
Cooley, W. W., 55
Cooper, R. G., 115
Coote, G. E., 66
Copic, M., 32
Cordoliani, M. L., 46
Cornman, W. R., 22, 28
Corvol, P., 129
Craddock, P., 70
Craig, H., 70
Craig, V., 70
Cramer, J. A., 87
Crawley, F. E. H., 119
Cress, L. W., 85
Crouzel, C., 79, 82, 107
Crowell, M. R., 70
Crum, L. W., 129
Crummett, J. G., 65
Csak, J., 26
Csikai, J., 29
Csom, G., 23
Cuello, O., 2
Cukor, P., 27
Curie, I., 73
Cutkova, O., 19
Czerwinski, A., 132
Czulak, C., 8

Dabek, T., 31
Daellenbach, C. B., 13
Dahlke, L. W., 26
Dakin, J. T., 109
Dakshinamurti, K., 125
Dalton, J. L., 19
Dams, R., 25
Dande, I. D., 32
Danhoffer, D. F., 17
Daniel, J. Y., 129
Dante, U., 2
Darrell, K. G., 113, 114
Das, H. A., 27, 67, 68
Das, N., 26
da Silva, C. P. G., 86
D'Auria, J. M., 37
Davidson, L. M., 127
Davidson, T. E., 62
Davies, B. H., 123
Davis, M. A., 105

Davison, C. C., 66
Davy, H., 33
De Bruin, M., 70
De Brun, J. L., 68
Decostre, P., 104
Deglume, Ch., 92
De Grazia, J. A., 77
Dejmokova, E., 19
De Jong, C. M. M., 129
De Jong, F. H., 129
De Kleijn, J. P., 87
De la Llosa, P., 79
De la Pena, A., 129
De Nardo, G. L., 93, 107
De Neef, J., 19
Dent, J. G., 116
De Soete, D., 36, 39
Detkova, N. V., 24
Deutsch, J. P., 92
Dewanjee, M. K., 105
Diamanti, C. I., 102
Dibbs, H. P., 23
Digenis, G. A., 87
Dimic, V., 32
Dimitroulas, C., 2
Dincer, S., 26
Dinwoodie, R., 79, 84
Dixon, J. E., 66
Dobrota, M., 121
Dodson, R. W., 35
Dolan, K. W., 26
Dolphin, D., 105
Donaldson, E. M., 129
Doran, J. E., 43, 51, 57
Dorgebray, G., 6
Dothan, M., 62
Dresia, H., 14
Drost-Wildschut, H., 107
Duarte, V., 26
Dubenskov, P. I., 13
Duffey, D., 26
Dumortier, A. G., 104
Dunn, A., 116
Dutson, T., 22
Dvorak, V., 103
Dyad'kin, I. G., 30
Dyck, J., 116
Dyer, A., 108
Dzuimkowski, B., 19

Eakhtiarov, A. V., 26
Eakins, J. D., 113, 119
Eckelman, W. C., 102, 104, 105
Egiazarov, B. G., 31
Egorov, E. V., 30
Eicholz, G. G., 22
Elagin, V. B., 23
El Garhy, M., 96
Elias, H., 96
El-Kady, A. A., 26
Ellem, K. A. O., 116
Ellis, W. K., 17, 21
Elsborg, L., 108
El Thawil, M., 10
Emeleus, V. M., 36, 67
Emel'yanov, V. A., 24
Emmelmann, K. P., 12
Emoto, Y., 35
Ender, M. L., 129
Endow, J. S., 105
Engh, T. A., 7, 8
England, B. G., 129
Engler, R. A., 82

Author Index

English, D., 104
Engstrom, M. A., 82
Erdal, B. R., 107
Erdmann, D. E., 27
Erdogan, H., 19
Eskes, J. M., 23
Evans, G. V., 4
Evans, H. J., 17
Evans, J. G., 129
Evans, R., 132
Exall, D. I., 8
Ezawa, O., 76

Fabre, L. F., jun., 129
Fahland, J., 25
Fairchild, R., 107
Fakhrvaezi, F., 25
Fallais, C., 103, 104
Fallot, P., 125
Farm, P. D., 26
Farmer, R. W., 129
Farrer, P. A., 100
Farrior, W. L., 25
Farsorra, C., 129
Favart, D., 92
Feather, S. W., 39
Fedorff, M., 26
Fel'dman, J. I., 31
Felici, M., 70
Fellows, C., 68
Feltkemp, T. E. W., 96
Ferguson, J., 70
Fields, P. R., 71
Fik, H., 9
Finn, R. D., 78, 89
Fiorelli, G., 129, 130
Firganek, H., 9
Firnau, G., 86, 87
Fischer, G. A., 129
Fischer, H., 16
Fisher, J. C., 17, 19
Fishman, M. J., 27
Flanagan, F. J., 41
Fleischer, M., 41
Fliegal, C., 105
Foley, D. H., 58
Follo, A., 19
Fomenko, I. N., 26
Fomenko, V. T., 26
Fookes, R. A., 17, 20
Forti, G., 130
Fortman, B. L., 102, 105
Foshag, W. F., 68
Fougea, D., 25
Fouqué, M., 34
Fourcy, A., 27
Fowler, J. S., 73, 82, 89, 96
Fox, B. W., 111, 121, 122
Frana, J., 26
Franchi, F., 129
Francis, W. H. G., 86
Frantz, A. G., 129
Fraser, H. M., 129
Frazzoli, F. V., 19
Freed, B. R., 107
Frevert, E., 23
Fricke, U., 122
Friedman, A. M., 45, 71, 107
Friedman, H. P., 51
Fries, B. A., 2. 10, 13
Fromageot, P., 79
Fugh, K., 115
Fujii, A., 100
Fulda, S. H. S., 26

Furater, T., 20
Furukawa, M., 107

Gagliardi, G., 129
Gangasharan, S., 26
Garcia Agudo, E., 2
Garnett, E. S., 86
Garnett, S., 87
Gasparrini, G., 32
Gavrilov, A. P., 11
Gazzard, T., 65
Geisler, R., 27
Gelbard, A. S., 76, 81, 84, 85
Genazzani, A. R., 129
Genunche, A., 87
George, J. H. B., 1
Gera, D. F., 30
Gerald, P., 25
Gerding, T. J., 19
Gerrard, M., 6
Gersey, F., 17
Gezing, M. J., 110, 116
Ghaleb, M., 25
Gholson, R. K., 119, 120
Giannotti, P., 129
Gibbons, D., 25
Gijbels, R., 26, 36, 39
Gillieson, A. H., 19
Gilson, A. J., 94
Gindler, J. E., 107
Girardi, F., 106
Givens, W. W., 25
Gjika, H., 130
Gladysz, C. V., 26
Glaeser, M., 12
Glass, H. I., 94
Glentworth, P., 90
Gloria, I., 79, 84
Goddard, R. E., 7
Godeau, A., 29
Goetz, L., 106
Gogan, F., 123
Gogan, P., 123
Goldstein, E., 84
Goldzieher, J. W., 129
Gomez, H. H., 2
Gomez-Sanchez, C., 129
Gomm, P. J., 119
Gonzalez, E., 127
Goodwin, D. A., 102, 105
Gorbachev, A. N., 26
Gordon, B. E., 113
Gordus, A. A., 38, 39, 61, 67
Gordus, J. P., 67
Gorev, A. V., 22
Goris, M. L., 94
Gorshkov, V. V., 26
Gorski, B., 2
Gorski, L., 25, 30
Gottschalk, A., 84
Goulding, R. W., 81, 82, 87, 88
Gountchev, C., 27
Graham, G. E., 82
Grahl, G., 2
Grant, P. M., 107
Granlund, R. W., 120
Gravitis, V. L., 17, 19, 20
Gray, F. C., 107
Graybill, F. A., 45
Greene, M. W., 107
Gregor, T., 15
Greig, R. A., 19
Grennes, R. A., 64

Griezer, F., 23
Griffin, J. B., 38
Grip, C.-E., 8
Groen, F. C. A., 70
Groothedde, M., 103
Grove, R. B., 102
Gruber, U., 17
Grudkin, K. A., 31
Grynszpan, R., 26
Guillaume, M., 93
Gunasekera, S. W., 88, 102
Gunn, A., 129
Gupta, R. N., 129

Haber, E., 130
Haccoun, A., 15
Hafez, M. B., 131
Hagat, L. P., 90
Hahn, G. J., 113
Hairland, R. T., 117
Hales, C. N., 130
Hall, A. W., 23, 25
Hall, E. T., 33, 62, 67
Hall, H. E., jun., 31
Halleman, D. F., 117
Halynska, B., 19
Hambright, P., 104
Hammersley, P. A. G., 100
Hammond, G. C. M., 113
Hammond, N., 68
Hamouda, I., 26
Handler, L. H., 132
Hansson, L., 7, 8
Hara, T., 76, 81, 107
Harada, S., 104
Harangazo, M., 10
Harbottle, G., 37, 38, 42, 44, 52, 64
Hardy, E. L., 17
Harper, P. V., 79, 84, 107
Harry, D. S., 119
Harvey, B. H., 113
Havronek, E., 19
Hayakawa, T., 127
Heard, M. J., 106
Hearst, J. R., 24
Heckman, R. V., 16
Hegedus, F., 93
Heinonen, J., 41
Heizer, R. F., 38, 69
Helf, S., 23
Hellwig, K. D., 22
Helman, E. Z., 110, 128
Helus, F., 86, 90, 98
Hendrix, W. A., 22
Hendry, G. O., 75
Henkelmann, R., 26, 27
Hennam, J. G., 129
Henrickson, E., 71
Herbland, A. M., 132
Herchl, M., 19
Hermann, H., 67
Hernandez, J. A., 46
Herscowitz, H. B., 128
Heslop, A., 30
Heslop, J. A., 1
Hester, T. R., 69
Hevesy, G., 34, 73
Hewitt, J. S., 32
Hichens, M., 130
Higashimura, T., 69
Hines, H. H., 93
Hinton, R. H., 121
Hinzpete, A., 122

Hirata, M., 107
Hnatowich, J., 99
Hodson, F. R., 43, 51, 57
Hoffer, P. B., 84
Hoffmann, B., 129
Hofner, R., 104
Hold, A. C., 13
Holder, J., 94
Holland, C. G., 69
Holloway, G. E., 27
Holloway, J. A., 32
Holymska, B., 19
Homewood, C. A., 109
Hope, H. J., 125
Hopkinson, E. C., 31
Horlock, P. L., 105
Horrocks, D. L., 108, 128, 131
Hortman, A. G., 79
Hoste, J., 26, 29, 36, 39
Hotte, C. E., 103
Howard, P. Y., 129
Howarth, W. J., 17, 19
Hoyans, A. F., 130
Hrdlicka, Z., 32
Hudson, R. F., 94
Hughes, M. J., 70
Hughes, W. L., 130
Hull, C. H., 60
Hupf, H. B., 94
Hurwitz, J. K., 26
Husak, V., 103

Ice, R. D., 103
Ido, T., 86
Ihlo, J. E., 101
Iio, M., 107
Ilic, R., 32
Inoue, Y., 26
Iofa, B. Z., 90
Isaacs, H. S., 104
Ishchenko, V. I., 31
Ishikawa, H., 109, 110, 114, 123
Ivanov, R., 11
Ivonov, V. M., 19
Iwamoto, M., 86
Iyer, R. S., 87
Izumnov, B. N., 20

Jackson, K., 17
Jacobson, L. A., 25
Jalnina, T., 25
James, R. W., 78
Janczyszyn, J., 30
Janshalt, A.-L., 107
Jasenak, V., 10
Jaumier, J. J., 27
Jeffcoate, S. L., 129
Jeghers, O. M., 104
Jenkins, J. G., 60
Jenkins, M., 63
Jester, W. A., 10
John, J., 29
Johnson, P., 1, 4, 8, 116
Johnson, R. B., 31
Johnsonbaugh, R. E., 116
Johnston, G. S., 100
Johnstone, C. W., 25
Joliot, F., 73
Jombik, I., 15, 19
Josza, I., 29
Joynt, R. C., 20
Jungerman, J. A., 93

Jungers, R. H., 19
Junzendorf, H., 19

Kaiser, D. G., 129
Kalbhen, D. A., 91
Kamphuis, J. A. A., 84
Kaplan, E., 107
Kaplan, N. M., 129
Karamanova, Zh., 20
Karasawa, T., 107
Karlstrom, K. I., 78
Karpunin, A. M., 26
Karten, F. H. S., 76
Kastner, M., 71
Kato, M., 16
Katorcha, G. P., 31
Katsurayama, K., 31
Katz, A., 52
Katzenmeir, G., 10
Kaucic, S., 98
Kawadam, D., 97
Keane, P. M., 129
Kehler, P., 13
Kem, D. C., 129
Kempi, V., 105
Kerimov, T. G., 30
Khamrakulov, K., 26
Kharkar, D. P., 66
Khentigan, A., 82
Kierzek, J., 9
Kim, Y. S., 2
Kim, S. Y., 130
Kimura, K., 23
Kinard, F. E., 111
Kinchenko, A. M., 31
Kinchenko, N. M., 22
Kinser, W. H., 11
Kirk, D., 59
Kirschner, A. S., 103
Kisama, K., 116
Kishpaugh, J., 59
Kitsenko, O. A., 24
Klas, J., 20
Klein, J., 108
Klein, R., 23
Kleinan, J. O., 116
Klempner, K. S., 20
Klimint, V., 25
Klovan, J. E., 59
Knight, L., 90
Ko, K., 129
Kobayashi, Y., 108
Koenders, E. B., 118
Kohl, P., 38, 69
Koike, J., 16
Koike, Y., 110
Kolaski, G., 9, 26
Komarov, S. G., 24
Kominami, G., 104
Koncheski, J. L., 23
Konoplev, Yu. V., 24
Kook, C. S., 87
Kopecky, P., 98, 99
Korkonosov, V. P., 11
Korthoven, P. J. M., 70
Koshelev, I. P., 31
Kosmowski, A., 1, 15
Kotel'nikov, V. V., 19
Kotov, P. T., 30
Kozlova, M. D., 100
Kraay, C. M., 67
Kramer, H. H., 101, 103, 107
Kramich, K. F., 22
Krampit, L. A., 19

Krapivskii, E. I., 22, 31
Krasil'nikov, B. N., 30
Krasnoshchekov, V. N., 13
Krauss, O., 86
Krautkramer, P., 16
Kraznoperov, V. A., 30
Kreft, A., 23
Kremer, L. N., 107
Kreyndlin, I. I., 13
Krishnamurthy, G. J., 105
Krishnamurthy, K., 2
Kritchevsky, D., 127
Krizek, H., 79, 84
Krohn, K. A., 90
Krumbein, W. C., 45
Krumbiegel, P., 110
Kruse, S. L., 105
Krylova, T. D., 26
Kubasik, N. P., 128
Kučera, E., 7
Kucera, J., 26
Kuiesi, J., 7
Kukharenko, N. K., 30
Kukula, F., 26
Kume, S., 107
Kumizuka, H., 132
Kuo, T. Y. T., 84
Kuoppamaki, R., 7
Kurek, R., 9
Kushelevsky, A. P., 75, 107
Kusubov, N., 91
Kyrein, H. J., 129

Lahat, A., 30
Lahmann, W., 122
Laine-Boszormenyi, M., 125
Lakhmyuk, V. M., 31
Lamb, J. F., 78, 82
Lambrecht, R. M., 73, 87, 89, 90, 93, 95, 96, 105
Lamson, M. L., 103
Lance, G. N., 50, 51
Lane, R. O., 78
Langone, J., 130
Lankosz, M., 19
Lapinki, N. P., 32
Larsen, P. O., 120
Larson, J. G., 25
Larson, S. M., 100
Laska, L., 30
Lathrop, K. A., 79, 84, 107
Laufer, P., 79
Laughlin, J. S., 74, 76, 81, 84, 85, 107
Laurence, D. J. R., 129
Laverlochere, J., 29
Laws, S., 84
Lebedev, V. E., 24
Lebowitz, E., 75, 76, 103, 107
Lechtman, H. N., 67
Lecote, J. L., 29
Lee, A. W., 94
Lee, B. H., 2
Lee, R. E., 19
Leeflang, H. P., 32
Lefferts, K. C., 67
Leikin, A. V., 31
Leman, E. P., 19
Lembares, N., 79, 84, 107
Leonov, P. E., 20
Leonowicz, L., 13
Lerner, J., 45
Leroy, J., 12
Leurs, C. J., 76, 107

Author Index

Levi, H., 34
Levin, V. I., 100
Levine, L., 130
Levushkii, Yu. A., 26
Lewis, J. T., 105
Lewis, T. T., 110
Lieberman, R., 114
Lima, F. W., 25
Lin, M. S., 104, 105
Lin, S. S., 96
Lin, T. H., 82
Lindner, L., 76, 86, 96, 107
Lines, H. H., jun., 107
Linfoot, J. A., 109
Linn, T. A., jun., 26
Lipshitz, D. A., 130
Lipsitz, E. L., 103
Litterscheidt, H., 9
Liu, T. Z., 127
Livingood, J. J., 35
Ljunggren, K., 5, 7
Lloyd, S. R., 128
Lobanov, C. M., 26
Loberg, M. D., 107
Locks, W. S., 15
Loewenthal, T. L., 105
Logan, J., 115
Lohnes, P. R., 55
Lohr, D., 9
Lontiadis, J., 2
Loska, L., 25
Lu, W. D., 27
Lucas, A., 33
Luckenbach, A. H., 69
Ludwick, J. D., 114
Luenonschloss, J., 1
Luise, M., 129
Lundan, A. I., 19
Lutfirakhmanova, L. B., 24
Lutz, G. J., 25, 29
Lyon, W. S., 36

Maass, R., 105
McAfee, J. G., 102, 105
McCabe, W. J., 2
McCallum, G. J., 66
McChing, R. W., 32
McDermott, E. A., 129
McDonald, J. M., 84, 129
MacDonald, N. S., 76, 84, 85, 100
McDowell, J. W., 131, 132
McFllistrem, M. T., 84, 86
McFarlane, J. C., 128
MacGregor, R. R., 82
McGuigan, J. E., 128
Machaj, B., 23
Machulla, H.-J., 79
McIntyre, N., 119
McKay, A. S., 31
McKenzie, R. M., 119, 120
McKerrell, H., 62
McKillip, T. W., 128
McKinlay, P. F., 30
McKlvean, J. M., 131
Macksey, J. A., 26
McKusick, K. A., 105
McLaughlin, A., 84
McLure, W. O., 121
McNeils, D. N., 114
McRae, J., 104
Mader, K., 13
Maeda, K., 100
Mahalanobis, P. C., 54

Mahon, W. M., 13
Maier-Borst, W., 86, 90, 98
Maki, Y., 25, 30
Malakhov, G. M., 17
Maliewska, M., 13
Malinin, A. B., 100
Mamuro, T., 26
Mancini, C., 19
Manfra, L., 70
Mangan, E. L., 16
Mangialoyo, M., 32
Mannelli, M., 129
Mantescu, C., 87
Marafante, E., 106
Marazano, C., 79, 81
Marche, P., 79
Marcu, G., 27
Mardh, S., 117
Mariani, G., 129
Markiewicz, W., 1, 13
Markowicz, A., 19
Marr, H. E., 19
Marsh, R. H., 25
Martin, B. T., 129
Martin, J. W., 25
Martin, M. J., 119
Martinelli, P., 19, 25
Mashiko, Y., 15
Masi, U., 70
Masilungan, B. A., 25, 30
Masini, S., 129
Massalski, J., 24, 31
Masuoka, D. T., 84
Matera, F., 129
Matsui, Y., 20
Matsumoto, M., 15
Matthews, J. D., 78
Mattila, O. P., 19
Matukanis, L. F., 26
Maudsley, D. V., 108
Mayorga, A., 84
Mayron, L. W., 107
Maziere, M., 79, 81
Meaburn, G. M., 107
Means, J. L., 82
Meares, C. F., 102
Mechanic, G. L., 118
Meeks, N., 115
Meeuwissen, H. J., 87
Meier, V. A., 26
Meille, E., 63
Meinken, G., 103, 104
Melent'ev, V. I., 26
Melvin, H., 27
Menard, J., 129
Menchini, G. F., 129
Menglekamp, B., 13
Merkle, R., 1
Merlini, M., 106
Merrick, M. V., 102
Merrill, J. C., 105
Merriman, J., 23
Metta, D., 71
Meyer, B. R., 98
Meyer, G. J., 97
Meyer, P. K., 129
Meyers, P., 35, 39, 42, 63, 67
Meyers, W. G., 78
Michalik, J. S., 7
Michel, H., 37, 61
Midgley, A. R., jun., 129
Mignard, J. L., 46
Mijagawa, K., 15
Milder, M. S., 100

Miles, L. E. M., 130
Miller, V. V., 26
Millon, R., 66
Mills, W. R., 23
Millstein, R., 87
Milsted, J., 45, 71
Mineski, R., 25
Mion, H., 129
Mishkin, F. S., 105
Misilek, L., 15
Misra, S. C., 26
Misztal, Z., 13
Mitchell, J. P., 32
Mitta, A. E. A., 76
Mizhata, A., 26
Mociomita, G., 26
Mock, B., 107
Moghissi, A. A., 114, 128
Moiseev, V. N., 31
Monahan, W. G., 84
Monna, D., 70
Moore, A. W., 52
More, R. D., 75
Morgan, D. W., 13
Morgat, J. L., 79
Mori, C., 16
Mornoi, T., 16
Moroshnichenko, V. M., 30
Moroz, N. I., 30
Morris, D. F. C., 26
Morstin, K., 24, 31
Moucka, V., 26
Mrowiec, J., 9
Mucha, R., 14
Mudrova, B., 98, 99
Muehldorf, V., 2, 8
Mueller, E. B., 112
Mukai, K., 26
Mukherjee, A. D., 27
Mukhin, S. S., 13
Muminov, V. A., 26
Munkov, B. Ya., 1
Murad, E., 19
Murakami, Y., 76
Muraoka, Y., 100
Murata, Y., 15
Muromtsev, Z. G., 31
Murphy, M. R. V., 119
Murray, H. C., 129
Murray, J., 33
Murrenhoff, A., 35
Musilek, L., 16
Myburgh, J. A., 97
Myers, W. G., 107
Myrin, N. Yu., 1

Nagorny, K., 25
Nahmias, C., 87
Nakahira, S., 8
Nakamura, H., 109
Naoum, M. M., 10
Nargolwalla, S. S., 29
Nastich, F. L., 31
Nath, A., 90
Navalikhin, L. V., 26
Nearne, K. D., 109
Neary, M. P., 110
Neely, H. H., 84, 100
Neirinckx, R. D., 96, 97
Nelson, D. E., 37
Nemeikova, A., 19
Neri, P., 129
Newton, D., 106
Newton, J. R., 129

Newton, R. G., 57, 65
Nichols, L. L., 32
Nicolini, J. O., 101
Nie, N. H., 60
Niemi, A., 7
Nissen, H.-U., 70
Niswender, G. D., 129
Noakes, J. E., 110
Noguera, E., 43
Nolan, P. F., 13
Norjiri, T., 25, 30
Norris, J. F., 125
Norton, E., 96
Notea, A., 11
Novikov, V. S., 13
Nowak, M., 9
Nozaki, T., 86, 107
Nugent, C. A., 129
Nunn, A. D., 90

Obeid, M., 25, 26
O'Brien, H. A., jun., 107
Ochkur, A. P., 19, 30
Oddy, W. A., 67
Oezyol, H., 26
Ohara, T., 2
Oldham, G., 27
Oldham, K. G., 119
Olin, J. S., 62, 64
Olivier, D. C., 48
Olsen, E., 71
Oniciu, L., 104
Op de Beeck, J., 26, 28
Orange, J. M., 25
Oselka, M. C., 92
Ostermann, P., 27
Ott, A., 15
Ovechkin, V. V., 26
Owers, M. J., 19
Owlya, A., 25
Owsiak, T., 19, 31
Ozoglu, I., 19

Paap, H. J., 31
Packer, T. W., 17, 19
Paic, G., 98
Painter, K., 110, 116
Palcos, M. C., 101
Palige, J., 7
Palino, G. F., 26
Palmer, A. J., 81, 82, 87
Palser, R., 103
Panaitesui, I., 32
Papez, K., 19, 23
Parfenov, P. T., 30
Parker, C. K., 13
Parker, R. P., 116, 117
Parker, W. C., 86
Parkhamenko, V. V., 20
Parkinson, M. J., 14
Parks, N. J., 84
Parsa, B., 25
Parsignault, D. R., 25
Parsons, L. A., 67
Passaglia, A. M., 25
Passoni, D., 32
Patek, P., 25
Patterson, C. C., 71
Patzer, R. G., 128
Pavlov, L. S., 14
Pawaskar, P. B., 29
Payne, K. W., 17
Pazmandi, L., 16
Pazzagli, M., 130

Peaker, F. W., 125
Pearson, J. D., 125
Peck, F. P., 22
Pedersen, L., 115
Pedley, R. B., 17, 22, 23
Peek, N. F., 84, 107
Peisach, M., 20
Pekarskii, G. Sh., 23
Perablo, R. A., 8
Perdijon, J., 29
Peresypkin, A. V., 11
Perlman, I., 37, 52, 61, 62, 69
Permour, P. H., 22, 29
Perris, A. G., 78
Perry, E. A., jun., 66
Persson, B. R. R., 105
Persyk, D. E., 110
Peto, G., 29
Petrosyan, L. G., 24
Petrov, P., 23
Pettit, W. A., 84
Phelps, M. E., 78, 107
Picon, M., 37, 46, 63
Pierce, T. B., 22, 26
Pindrus, P., 35
Pineri, M., 25
Piper, T. C., 13
Pirovano, B., 32
Pittman, K. A., 129
Pitts, R. W., jun., 30, 31
Plotnikov, R. I., 26
Pogell, B. M., 113
Poggenburg, J. K., 75
Pogue, J. E., 68
Pohl, K. P., 2
Poitou, J., 109
Polkovnikov, V. K., 11
Polyanchenko, A. L., 25
Ponomarev, V. S., 26
Popplewell, D. S., 113
Porada, E., 69
Postelnicu, C., 26
Postnikov, V. I., 1
Poston, A. M., 25
Potapova, Z. M., 100
Potsaid, M. S., 105
Pouraghabagher, A. R., 25
Pozdeev, D. B., 13
Pozdnikov, V. N., 11
Prag, A. J. N. W., 62
Prasad, K. N., 10
Pravikov, A. A., 13
Preston, D. F., 86
Price, L. W., 108
Prieels, R., 92
Priess, K., 15, 30
Profio, A. C., 25
Protti, D. J., 125
Przybylowicz, E. P., 29
Pyrovokakis, J. A., 119

Racataian, I., 26
Radicella, R., 101
Radin, N. S., 110
Radisov, E. M., 26
Radiwan, M., 7
Radzikowski, Z., 9
Rakovic, M., 25
Rakovskii, E. E., 26
Ramette, R., 71
Randolph, R., 117
Rands, B. C., 64
Rands, R. L., 64, 68
Ranschenbach, P., 123

Rant, J., 32
Rao, P. N., 129
Rao, S. M., 2
Rapkin, E., 108, 115
Rauschenbach, R., 115
Ravetz, A., 67
Rawley, T. B., 19
Raymond, C., 79
Reba, R. C., 102
Reddy, G. R., 29
Redgewell, R. J., 120
Redvanly, C. S., 95
Reed, D. J., 19
Reed, M. E., 84, 86, 87
Rees, A. F., 123
Regunato, R. J., 12
Reid, J. V., 25
Reide, F., 109
Reiffers, S., 84
Reinig, W. C., 22, 29
Remick, F. J., 10
Renaud, B. C., 23
Renfrew, C., 57, 66, 70
Resmini, F., 92, 106
Resnekov, L., 84
Resvanov, R. A., 22
Reudinger, A., 10
Reunanan, M. A., 123
Rey, S., 2
Rhodes, B. A., 104
Rhodes, D. F., 5
Rhodes, J. R., 19
Ricci, E., 36
Riceberg, L. J., 130
Rich, B., 79, 107
Richards, E. E., 36
Richards, P., 75, 76, 103, 104, 105
Richards, T. W., 34
Richardson, P. J., 114
Riedlmayer, L., 2
Riesing, J., 2
Riess, W., 126
Riffl, F., 29
Rijskamp, A., 84
Riviero, J. F., 129
Robbins, P. J., 102, 105
Robert, A., 19
Roberts, G., 67
Robinson, A. G., 129
Robinson, C. D., jun., 105
Robinson, G. D., 86, 87, 94
Robinson, R., 31
Robitaille, H. A., 32
Roctzer, H., 8
Rodden, A. F., 77
Roffmann, S., 127
Rogers, V. C., 26
Rohlf, F. J., 46, 48, 59
Rojkind, M., 127
Roman, M., 87
Rosentreich, M., 102
Ross, H. H., 36
Rossler, K., 96, 97
Rothenberg, S. P., 130
Rothwell, E., 13
Roux, H., 119
Rouxel, C., 68
Rowland, F. S., 87
Rowse, O. J., 2
Royer, M. E., 129
Rozental, J. J., 32
Rubin, J., 51
Rudenko, V. S., 26

Author Index

Runge, K., 2
Russell, J. S., 52
Rybach, L., 70
Rychvarov, A., 20
Rydin, R. A., 82

Sabbione, E., 106
Sacelean, V., 20
Saha, G. B., 100
Said, M., 8
Saied, F. I. A., 131
Sainte-Laudy, J. L., 79
Salmon, J.-M., 111
Sal'tsevich, V. B., 22, 31
Samadi, A. A., 26
Sanchez, W., 2, 10
Sandford, P. A., 119
Sankaran, L., 113
Sankar Das, M., 29
Sanquist, G. M., 26
Santoliquido, P. M., 26
Santos, G., jun., 26
Santos, M. E., 19
Sapakoglu, A., 19
Sargent, T., 91
Sato, O., 16
Sauti, S., 19
Savinkin, P. T., 30
Sayre, E. V., 35, 36, 37, 38, 39, 41, 42, 44, 55, 61, 62, 64, 65, 67
Sazonov, O. L., 11
Scales, B., 125
Scasnar, V., 25
Schaal, D., 27
Schaffer, F. L., 109
Schauceil, S., 128
Schimmel, A., 76
Schmidt, H. 110
Schmitt B. F., 26
Schneider, J., 27
Schnizleiu, J. G., 19
Scholz, K. L., 107
Schrier, B. K., 119
Schubiger, P. A., 39, 62
Schuelken, H., 32
Schult, O., 91
Schultz, W. E., 30
Schuster, N. A., 24
Schutte, L., 118
Schwartz, J., 84
Schweizer, F., 62
Sciutti, S., 32
Scott, J. H., 31
Seaborg, G. T., 35
Seda, J., 16
Segal, Y., 11
Segebade, C., 25
Seidel, A., 113
Seignemartin, C. L., 26
Seim, H. J., 26
Seki, R., 107
Sekowski, S., 16
Sel'dyakov, Yu. P., 31
Semimli, H., 26
Senko-Bulatniji, I. N., 31
Serafini, A. N., 94
Serio, M., 129, 130
Serva, L., 70
Servian, J. L., 75
Sevastianova, A. S., 100
Sevast'yanov, Yu. G., 90
Shalgosky, H. L., 19
Shani, G., 19

Shapovalov, E. P., 11
Sharabarin, V. D., 31
Sharpe, S. E., 111
Shastova, L. S., 30
Shelke, R., 23
Shelton, W. L., 32
Shenbarg, C., 20
Shepelev, G. J., 31
Sheppard, C. W., 1
Sherman, H., 24
Shih, H., 129
Shimilevich, Yu. S., 26
Shimose, S. T., 93
Shinkarchuk, V. L., 26
Shipley, B. A., 104
Shishakin, O. V., 31
Shlykov, V. S., 24
Shohet, S. B., 127
Sholz, K. L., 92
Short, M. D., 94
Shreiber, C., 130
Shulgin, A. T., 91
Siemsen, J. K., 97
Sieveking, G. de G., 70
Siewierski, E., 26
Siewierski, J., 9
Sigal, P. M., 14
Sigleo, A. C., 40
Signarbieux, C., 109
Silberstein, E. B., 102
Silvester, D. J., 75, 94
Simon, H., 115, 123
Simonesu, L., 87
Simpson, H., 1
Sinitsyna, T. S., 26
Sins, H. E., 128
Sirca, F., 32
Skidmore, M. R., 98
Slater, D. N., 65
Sliwowski, J., 125
Slobodic, M. J., 120
Smakhtin, L. A., 26
Smit, J. A., 96, 97
Smith, C. S., 72
Smith, E. W., 132
Smith, H. D., 30
Smith, H. S. P., 107
Smith, M. P., 24
Smith, Ph. B., 67
Smith, P. H. S., 105
Smith, R. W., 36, 65
Smith, T. D., 105
Smith, T. W., 130
Smoak, W. M., 94
Sneath, P. H. A., 45, 46, 50, 52, 58, 60
Sobkowski, J., 132
Sodd, V. J., 91, 92, 107
Sode, J., 116
Soergel, M. E., 109
Soito, N., 23
Sokal, R. R., 45, 46, 48, 50, 52, 58, 60
Sokolov, E. A., 30
Sokolov, M. S., 31
Soloman, H., 54
Sommerville, I. F., 129
Sorantin, R., 25
Sotskij, A. R., 17
Soulas, D., 129
Souri, E. J., 123
Southard, G. C., 68
Souto, H., 16
Sowerby, B. D., 19, 21

Spackman, R., 4
Spang, R. C., 11
Sparmhake, N., 120
Spaulding, J. D., 110
Specht, S., 27
Spiehler, V., 110
Spodenkiewicz, T., 13
Spolter, L., 84, 85
Spomer, B., 90
Springer Peacy, J., 70
Srebrodol'skii, D. M., 24
Srinavasan, V., 124
Staerk, H., 25, 27
Stames, P. E., 20
Stamn, W. J., 32
Stanef, I., 109
Stang, L. G., Jun., 75
Stanley, R. E., 128
Stanley Wood, N. G., 13
Starikov, N., 30
Stark, V. J., 84
Starzec, A., 24, 31
Stauffer, H., 91
Steeg, M., 16
Stehno, G., 23
Steigman, J., 103, 104
Steinberg, D., 115
Steinbrenner, K., 60
Steindler, M. J., 19
Steiner, A. L., 130
Steinnes, E., 26, 27
Stene, E. W., 1
Stenico, A., 63
Stevens, U., 129
Stevenson, D. P., 37, 38
Stewart, C. C., 17
Stewart, R. F., 23, 25
Steyn, J., 98, 114
Steytler, J. G., 115
Stöcklin, G., 79, 91, 96, 97
Stollar, D., 130
Stone, J. F. S., 57
Storelli, L., 19
Stott, A. N. B., 106
Straatman, M. G., 79, 81, 84
Stradling, G. N., 113
Straub, W. A., 26
Streuli, F., 107
Stross, F. H., 37, 38, 69
Stuart, J., 129
Sturkol, E. W., 16
Subramanian, G., 102, 105
Sucr, T. H. G. A., 76, 86
Sumiya, Y., 31
Sundberg, M. W., 102
Suschny, O., 41
Sutton, G. A., 113
Suzeu, H., 15
Svitel, J., 19
Swick, W. A., 19
Świgón, K., 7
Swinth, K. L., 32
Switsur, V. R., 115
Szabo, L., 26

Tabor, P., 16
Taddeucci, A., 70
Takada, K., 26
Takahashi, J., 84, 85, 100
Takano, T., 26
Takine, M., 110, 123
Takita, T., 100
Takiue, M., 114
Taksar, I. M., 11

Talanin, N. Yu., 26
Tamminen, A., 7
Tamura, N., 25, 26
Tan, M., 19
Tanaka, H., 104
Tanner, H. L., 31
Tanner, J. T., 27
Taushkanov, A. P., 31
Taute, W. J., 8
Taxar, I. M., 11
Taylor, D. M., 100, 107
Taylor, G. M., 16
Taylor, L. J., 1
Teben'kov, A. A., 31
Ten Haaf, F. E. L., 123
Terekhin, L. N., 14
Terent'ev, Eh. P., 27
Teresi, J. D., 77
Terrani, M., 63
Tesnavs, E. R., 11
Thakur, M. L., 90, 94, 100, 102, 105
Thomas, F. D., 105
Thomas, L. C., 57
Thorneycroft, I. H., 129
Thorngate, J. H., 131
Tibbetts, B. L., 31
Tilbury, R. S., 74, 76, 81, 84, 107
Ting, P., 128
Tittman, J., 24, 30
Tobia, S. K., 44
Tolgyessy, J., 10
Tong, J. Y., 78
Tornau, W., 96
Torok, I., 8
Traikov, I., 11
Trajkov, L., 1
Trap, A. E., 20
Trauter, J., 15
Tree, M., 129
Treves, S., 105
Troll, W., 127
Troughton, H., 75
Tschroots, R. J. M., 96
Tschurlovits, M., 131
Tseitlin, V. G., 22, 24
Tsujimoto, T., 31
Tubis, M., 105
Turekian, K. K., 52, 66
Turi, B., 70
Turner, N. A., 120
Tykva, R., 109
Tyler, F. H., 129
Tyler, J. F. C., 113, 114
Tyler, J. P. P., 129
Tyukaev, Yu, V., 30

Udenfriend, S., 127
Ulick, S., 129
Ulrich, H., 2
Umezawa, H., 100
Unfried, E., 131
Urbanski, T. S., 26
Urch, D. S., 90
Utt, O. L., 17

Vaalburg, W., 81, 84
Vagonov, P. A., 26
Valent, J., 22
Valk, P. E., 104
Valkovic, V., 19
van Dalen, A., 8
Vandercasteele, C., 26

Vandergraaf, T. T., 27
van der Kley, N., 32
van der Laarse, J. D., 121
van der Molen, H. J., 129
van'der Sloot, H. A., 27
Vandlek, T., 25
Van Dyke, K., 116
Van Grieken, R., 26
Vaninbroukx, R., 109
Van Vunakis, H. V., 130
van Zanten, B., 87
Van Zelst, L., 39, 63, 67
Varaljai, M., 16
Varsano-Aharon, N., 129
Varga, S., 10
Varvarskii, E. V., 31
Vary, A., 32
Vasil'ev, S. I., 30
Vasilev, V., 11
Vaswani, N. K., 23
Vaughan, M., 115
Veen, H., 125
Veenboer, J. Th., 76
Vehida, Y., 15
Vekic, B., 98
Velyus, L. M., 26
Verdonck, L., 129
Verhassel, J. P., 118
Verjans, H. L., 129
Vermeulen, A., 129
Vernon, P., 94
Verot, J. L., 27, 29
Viallet, P., 111
Vichy, M., 46, 63
Vieczorek, E., 26
Vikari, J., 127
Villa, M., 92
Villani, S., 63
Vinhlong, K. D., 29
Viol, G., 129
Virish, V., 26
Visser, J., 76, 107
Vlasyuga, S. P., 26
Vlatkovic, M., 98
Vogg, H., 25
Voigt, A. F., 71
Volat, J. P., 12
Vol'fshtein, P. M., 30
Voljin, V., 19
Volokh, V. A., 26
Vondruska, V., 32
von Lehmden, D. J., 19, 26
Voropaev, Yu. A., 13
Vorsatz, B., 26
Vuimova, E. S., 25

Wada, N., 23
Waechter, K. H., 15
Wagner, A., 126
Waite, M. A., 129
Wakakuma, S. I., 100
Wall, W. B., 25
Walter, H. E., 127
Walz, D. R., 77
Warashina, T., 69
Warbick, A., 105
Ward, G. K., 58
Warren, S. E., 65
Wasilewska, M., 19
Waters, S. L., 90
Watonabe, T., 16
Watson, I. A., 105
Watson, P. R., 119
Watt, J. S., 17, 19, 20

Waxman, A. D., 97
Waxman, S., 130
Weaver, J. R., 26, 37
Webster, S. L., 23
Wegner, L. A., 115
Weick, C. F., 35
Weigand, P., 38
Weill, A. P., 35
Weinreich, R., 91
Welch, M. J., 79, 81, 84, 90, 101, 107
Welch, T. J., 101
Wells, A. C., 106
Wenk, G. J., 19
West, C. D., 129
Wetter, L. R., 116
Whiston, J., 4, 8
White, N., 129
Whitehead, N. E., 66
Whittaker, E. I., 114
Widemann, F., 37
Widner, P. J., 82
Wiechowski, A., 132
Wieland, B. W., 78
Wier, K. E., 16
Wigfield, D. C., 124
Wiggins, P. F., 26
Wikjord, A. G., 27
Wilkins, S. R., 93
Wilkinson, L. R., 19
Williams, A. W., 19
Williams, H. P., 103
Williams, J. Ll. W., 62
Williams, K. F., 13
Williams, R. L., 87
Williams, T. A., 129
Williams, W. T., 50, 51
Wilson, H. H., 25
Wilson, R., 64
Wilson, R. H., 119
Wilson, S. H., 119
Wilson, T. B., 98
Winchell, H. S., 78, 82
Winder, F. G., 124
Winkelmann, H., 115
Winkler, M. A., 13
Winsor, P. A., 121
Winstead, M. B., 78, 82
Wojciechowski, A., 13
Woldring, M. G., 81, 84
Wolf, A. P., 73, 78, 83, 87, 89, 90, 93, 95, 96, 105
Wolf, M., 131
Wolf, W., 97
Womer, H. K., 8
Wong, C. H., 127
Wong, D. W., 105
Wood, D. E., 28
Wood, R. A., 100
Woodard, H. Q., 107
Woodhead, J. W., 130
Woods, M. J., 90
Wozniakowska, G., 125
Wright, G. A., 38
Wu, S. C., 27
Wyld, G., 37
Wyttenbach, A., 39, 67

Yakubson, K. I., 31
Yakubson, K. J., 25
Yamamoto, T., 25
Yamashita, T., 16
Yanehevskii, Yu. P., 19, 30
Yano, Y., 76

Author Index

Yanuskowskii, V. A., 11
Yeats, D. B., 105
Yegnasubramanian, S., 26
Yokoyama, A., 104
Yoshimoto, T., 31
Yoshioka, T., 100
Youmans, A. H., 31
Young, W. J., 67
Youssefnejadian, E., 129
Yuasa, T., 109

Yudin, V. A., 24
Yulo, M., 26

Zaghoul, R., 25, 26
Zapletal, L., 26
Zemplen-Papp, E., 26
Zhukovskii, A. N., 26
Zietlow, K., 25
Zitnansky, B., 22, 26

Zlatarov, V., 11
Zlotova, I. M., 25
Zlrick, R. H., 116
Zmijewska, W., 27
Zold, E., 16
Zonderhuis, J., 67
Zolle, I., 104
Zucchelli, G. C., 129
Zwittlinger, H., 26

QD
601
A1
R285
v.3
1974-75

NOV 13 1978